THE LINK

THE LINK

UNCOVERING OUR
EARLIEST ANCESTOR

COLIN TUDGE with
JOSH YOUNG

LITTLE, BROWN AND COMPANY
NEW YORK BOSTON LONDON

Little, Brown and Company
Hachette Book Group
237 Park Avenue, New York, NY 10017
Visit our Web site at www.HachetteBookGroup.com

First Edition: May 2009

All photographs copyright of Atlantic Productions, all X-rays copyright of
Jørn Hurum, University of Oslo, and all illustrations copyright of Zoo.

Little, Brown and Company is a division of Hachette Book Group, Inc. The
Little, Brown name and logo are trademarks of Hachette Book Group, Inc.

ISBN 978-0-316-07008-9

10 9 8 7 6 5 4 3 2 1

RRD-IN

Printed in the United States of America

To Ida

CONTENTS

FOREWORD

When I first saw an image of this remarkable fossil, I couldn't sleep for two nights. I never imagined that I would see and be able to work on such a unique specimen. It has truly been a once-in-a-lifetime experience.

Without the foresight of the museum director, Professor Elen Roaldset, and the support of the board of the Natural History Museum at Oslo University, none of this would have been possible. Maybe this specimen would even have ended up in another private collection—unseen by the scientific community and unknown to the world at large.

I feel privileged to have been the first scientist to describe the specimen and to have been able to collaborate with some of the world's best scientists in the field. Jens Franzen of the Senckenberg Research Institute in Frankfurt has worked tirelessly with me to describe the specimen's anatomy. Without Jörg Habersetzer's cutting-edge X-rays and CT scans, we would not have been able to draw the detailed conclusions that we have. Wighart von Koenigswald of the Steinmann Institute for Geology, Mineralogy, and Paleontology at the University of Bonn described the specimen's preservation, and Philip Gingerich and Holly Smith of the Museum of Paleontology and Museum of Anthropology respectively, at the University of Michigan, Ann Arbor, have been invaluable with their deep understanding of primate anatomy and especially dentition. In building my dream team, I have met some incredible academics and made lifelong friends.

But this journey has not been only an academic one. When one works on a primate, and a primate that could be a distant ancestor of ours, one can't help forming a personal bond with the specimen. This fossil will stay with me for the rest of my life—a snapshot in time of a young girl's life. It often reminds me of my own little girl growing up at home, and that is why I came to name the fossil after her—Ida.

The authors of this book, Colin Tudge and Josh Young, have described our research and the wider story of Ida beautifully. It is an unfolding tale that I hope will capture the readers' imagination of this period of our own evolution.

A special thanks to Anthony Geffen, the founder and creative director of Atlantic Productions, London, and his documentary team, who realized the possibilities in Ida and managed to tell her story to the world.

—Jørn Hurum, 2009

THE LINK

CHAPTER ONE

DISCOVERING IDA

In the glow of a gibbous moon, a petite being moves through the palm trees surrounding a lake that seems almost impossibly pristine. The small creature living in this lush tropical forest has a light coat of fur, and she's less than two feet (half a meter) tall. With forward-facing eyes, her elongated head is slightly overproportioned to her body, suggesting intelligence. Her legs are a bit longer than her arms, allowing her to climb trees and move between them to avoid dangers on the ground below.

This is Ida, less than a year old and just weaned from her mother. She now has the freedom to roam, climb, and fend for herself. Moving as though she's chasing down the wind, she pushes off one branch with her feet, using her tail like a rudder to guide her, and then grasps the next branch with her long fingers. She secures her position with her toes, all of which are of nearly equal length and used for nothing except movement. Her opposable thumbs enable her to gracefully grab and go.

As she searches for her next meal, Ida ignores the varieties of insects, easy targets all, and settles on a piece of fruit. She wraps her hands around the fruit, pulls it from its branch, and pops it into her oblong mouth. Moving her jaw rhythmically, she chews the fruit with her rounded teeth. For a living being, searching for food in any forest would usually be a rather straightforward process—were it not happening 47 million years ago in this particular forest, by this particular lake.

The rain forest where Ida lives would be recognizable to us but not identical to any we have seen. It is a sight to behold, and though its features are relatively common for its time, it's a place that could only have been made by a special confluence of events. It has a warm and equitable climate that stimulates the growth of its plants and trees, making life comfortable for its inhabitants. Palm trees with large root masses shoot into the air, as do cycads, which have a stout trunk at their base and large, tightly wound bundles of leaves at the top. Pygmy horses prance the verdant land. Opossums and armadillos share space with giant mice and salamanders. Birds with woodpecker-like beaks and short wings fill the air, but there are also powerful flightless ground-dwelling birds six feet (almost 2 meters) tall that are feasting on mammals. Large insects protect themselves by imitating tree leaves. Rat-size "longfingers" with tails twice the length of their bodies rip bark from the trees with their two clawlike fingers in hopes of finding insect larvae. Anteaters seethe as they eye giant ants an inch (2.5 centimeters) long, but these ants often make last-second getaways by spanning their wings a full six inches (15 centimeters) and beating a retreat.

In the center of the forest sits the lake, which is an endless source of fascination for the creatures living around it.

The crater it fills was formed by a volcanic eruption thousands of years before Ida was born. Deep underground, the earth's crust split, sending boiling-hot molten magma rushing toward the surface. Just before the magma broke through the surface, it collided with a layer of groundwater and instantly turned into steam, a process that eliminated all the lava. This reaction caused an explosion that blew out a large chunk of earth and left behind a vast crater a mile (1.6 kilometers) wide and more than eight hundred feet (250 meters) deep known as a *maar lake*.

Over time, the hole filled with a combination of groundwater seeping from below and rainwater trickling down from the heavens, and a lake was created. Though there are a few streams, no rivers flow in and out, so the water in the lake remains relatively still. Because of the lack of currents, the water at the bottom is cut off from the upper layers and unable to draw down oxygen from the atmosphere above. All the fish live near the top, and there are no scavengers prowling the lake floor.

The lake is so rich in algae that from above, it resembles a green eye in the middle of the rain forest. As the surface algae die, they sink to the bottom and turn to slime. Eventually the slime turns to mud. The combination of this heavy mud and an almost total lack of oxygen kills nearly all the bacteria, allowing any creature that perishes and sinks to the bottom to rest virtually undisturbed for eternity.

The lake is the heart of this forest ecosystem, and it sustains a diverse array of life. At the water's edge, enormous crocodiles patrol their territory, and frogs make a *tock* sound as they search for insects. The land frogs have short hind legs and dig for food, while the water frogs have long, thin legs. In the water, turtles with paddle-like feet push their way across the surface of the slimy water. Shells are attached to sunken rocks lining the sides of the lake. Repelled by the lack of oxygen at the deeper levels of the lake, the bowfin, perch, gar, and eels swim near the surface. Many of the fish eat hard-shelled prey such as snails.

As Ida moves through this vast array of wildlife, dodging bats and staying out of reach of the saw-toothed crocodile's snapping jaws, it's clear that she is different from the others resembling her. She's much lower in the trees than they are, and at first she appears to be playing with the other wildlife. But as she moves, it's apparent that she's tentative with her right wrist and her left arm.

Amid the tranquillity of this Eden, and the chattering of hundreds of animals and the sounds of a squealing flock of bats, a rumble begins from beneath the lake and quickly becomes a roar, but the animals in the surrounding forest are oblivious to the large gas bubble erupting from deep within the earth's crust.

At that moment, Ida is leaning down for a drink from the lake, grasping the shaft of a palm tree with one hand and reaching to cup the water with the other. She seems apathetic about the disruption.

The perfectly shaped oval bubble races up through hundreds of feet of water, killing almost instantly everything swimming in it, and finally releases a thin layer of dense gas as it breaks the surface. The gas is heavier than ordinary air, so it clings to the ground, covering the water's surface and creeping across the low-lying ground.

Ida detects the malodorous fumes. All the creatures do. Like them, she reacts immediately, but her arm is not strong enough to quickly pull herself up and away from the water's edge. The gas engulfs her airspace. She doubles over into a fetal position, slips into unconsciousness, and collapses into the water along with all the others in the vicinity.

Lifeless, Ida sinks to the bottom of the lake and comes to rest in the mud. The natural lifecycle of one young being is complete. But because of the wondrous oddities that make up the time, place, and circumstances of Ida's death, she might leave an indelible mark on history, more than any being that lived within millions of years of her.

Forty-seven million years later, Earth has changed. The Indian tectonic plate has collided with the Asian plate, resulting in the formation of the Himalayas. The polar ice caps have formed. The modern continents have taken shape, and climate change has occurred—many times. Humans have evolved. And, in a relative blink of time, the modern history of man—the development of civilization, the agricultural, industrial, and technical revolutions, and the fighting of wars—has occurred. All this as Ida lay still in the earth.

Today's Earth looks very similar to the Eocene Earth, but the two are not identical. The shifting of the continents over millions of years has moved the tropical forest's crater lake 150 feet (45 meters) beneath the Earth's surface and relocated it from what is now the Mediterranean Sea in the area of Sicily to about twenty-two miles (35 kilometers) southeast of Frankfurt, Germany, near the village of Messel. During this time, the weight of the thick mud has compacted the layers of dead algae into an oily shale and flattened the remains of thousands of creatures that died there, including Ida.

The oily shale deposits were discovered by coal prospectors

in the eighteenth century, and the quarry, now named the Messel Pit, became the site of furious activity when a process to convert the shale into raw petroleum was perfected. On December 30, 1875, the Messel Pit provided its first hint that something special was buried there when the bones and jaw fragments of a crocodile were found. Even as there were fossil recovery missions on an ad hoc basis through the 1900s, the mining continued unabated. Somehow, in all the frenetic digging, Ida's remains were missed.

Finally, in 1966, formal Messel Pit excavations were undertaken by paleontologists and archaeologists. Fossils of horses, fish, bats, and crocodiles perfectly frozen in time were unearthed and preserved. In many cases, complete skeletons were preserved, along with bacterial imprints of hair, feathers, scales, and even internal organs. But the discovery of the first primitive horse caused day-tripping fossil hunters armed with rock hammers, wire brushes, and tiny cleaning towels to ravage the pit in search of keepsakes or that rare find that they could sell on the open market. Layer by layer, tons of shale were removed, and the depth of the pit was soon almost two hundred feet (60 meters).

By 1971, mining had ceased, crippled by competition from cheap oil imports. That year, the German government, which wasn't benefiting financially from the fossil expeditions, decided the pit was an eyesore and declared that it would become a landfill. An access road was even constructed, but the scientific community protested and launched an all-out campaign to save the historic site. The protest delayed plans for a landfill and left the pit open to fossil hunters until a resolution could be reached. Messel's possible destruction sparked a fossil rush. Both scientists and collectors tried to haul as many fossils as they could out of the pit as quickly as possible, and it was a case of finders keepers.

Sometime in 1982, on a routine day for most of the world, a man living on the outskirts of Frankfurt went on a day expedition to the pit to add to his private fossil collection. Unlike those who had pillaged the sacred quarry, he took great care in his work and preserved each fossil on site like a scientist.

While splitting the layers of shale, the fossil hunter stumbled on a fossil of what looked like an exotic monkey crushed to the thickness of a silver dollar. It was Ida, frozen in a fetal position, exactly as she had come to rest on the bottom of the lake.

He realized that he was onto something. He carefully extracted the fossil from the ground and diligently wrapped it in wet newspaper. He then returned to his house and probably employed an expert to prepare the fossil—its preparation is so skillful that just a handful of people in the world could have done it. It must have taken months of careful chipping away the shale and stabilizing the bones before the specimen was ready to be placed on a shelf in his basement with the others he had excavated, away from the eyes of science and the public, for him alone to see.

Fossil collectors can be an obsessive bunch, much like art collectors who hoard famous masterpieces. Many of them merely want to own precious world treasures without telling anyone that they have them in their possession. Some want to withhold them from scientific study. Others are competing for attention in an underground society. For whatever reason, the private collector who found Ida's fossilized remains simply put her on a shelf for the next twenty-five years.

Roughly 250 miles (400 kilometers) north of the Messel Pit, the second-largest fossil fair in Europe is held each December in Hamburg, Germany. In 2006, thousands of people attended, and dealers from all over the world peddled their wares. The fair attracts a diverse crowd. Scientists in tweed sports jackets with elbow patches look for specimens for their museums. Private collectors prowl for that one special snapshot of time. Dealers look for items that can be sold on the black market. And, because of the time of the year, locals shop for unique Christmas presents.

The fair takes place in an exhibition hall that covers half a square mile. Rows of tables display polished stones, diamond crystals, and fossilized animal parts dangling from necklaces. Wall-size plates of rock with imprints of exotic fish are propped up for viewing. To the untrained eye, the fair looks like nothing more than table upon table of rocks. To the trained eye, sometimes that's exactly what it is. Often the best specimens are not on display but rather held by dealers under the tables or in their cars for those who will truly appreciate them—and for those who will pay.

Jørn Hurum, associate professor of paleontology at the Natural History Museum at the University of Oslo, is a regular at the Hamburg fair, where he goes each year in hopes of adding to his museum's collections. He has traveled the world, looking for connections between species. At age forty-one, he has the rugged look of an explorer who spends a lot of time in remote areas of the planet. His long hair is pulled over his forehead, and he has a sturdy build. His bright eyes reveal a childlike enthusiasm for his craft that dates from his youth.

Hurum grew up outside Oslo, and from the age of six he knew that he wanted to be a paleontologist. The moment came when his parents were reading him a story about a boy who was walking on the shore, throwing stones into the sea. In the story, one of the stones says to the boy, "Don't throw me into the sea. I'm a fossil. I can tell you a story." The stone, a 500-million-year-old trilobite that existed before fish and dinosaurs, starts to tell the boy about how life evolved over millions of years. Hurum was so taken by the story that he wanted to know everything about fossils, and a lifelong passion emerged.

Hurum studied paleontology at the University of Oslo as an undergraduate, earned his PhD in 1997, and has since become an associate professor in vertebrate paleontology at the university. In 2006, Hurum led a team that described the fossil of a plateosaurus buried more than one mile (1.6 kilometers) beneath the earth's surface. The dinosaur, estimated to have been thirty feet (9 meters) long and weighing four tons, lived 200 million years ago, during the Triassic period. His later fieldwork in the Svalbard archipelago, located in the Arctic Ocean midway between Norway and the North Pole, resulted in the mapping of forty skeletons of Jurassic marine reptiles and the excavation of six skeletons. His team's greatest find was a large 150-million-year-old sea reptile. This fifty-foot (15-meter) pliosaur, dubbed "the Monster," is the longest pliosaur known to science, with a body of forty feet (12 meters) and a skull measuring ten feet (3 meters).

When Hurum arrived at the Hamburg fair in December 2006, he had no idea that this routine trip would change his life. Early one afternoon, Hurum and his museum colleague, Dr. Hans Arne Nakrem, were milling around the table of a reputable dealer named Thomas Perner. Hurum had bought several small specimens from Perner over the years and had developed a working relationship with him. But when Hurum caught sight of Perner, he noticed the dealer was staring at him and acting strangely. To Hurum, Perner looked like a man burdened with a secret that he needed to reveal.

Finally Perner approached Hurum and whispered, "I need to show you something so interesting and unbelievable, but there are too many people here at the moment. Can I buy you a drink at four o'clock?"

Hurum immediately accepted. Perner was very reliable, so Hurum was certain he had a worthwhile specimen to show him. But why the secrecy? he wondered. He presumed it was a Messel specimen, possibly of a horse.

Hurum and Nakrem returned to Perner's table at four P.M., and the three of them walked to a small bar inside the exhibition hall. The bar's specialty was fresh-squeezed fruit juice, and Perner ordered three mixed juice and vodkas.

Surrounded by temporary exhibition curtains and attendees wearing name tags, Perner explained to Hurum that a private collector who demanded anonymity had given him six months to sell the fossil he was about to show him. The collector was getting on in years, and he wanted to remain unknown so he wouldn't be harassed for not having put it into the scientific world earlier. Perner then opened an envelope and pulled out a high-resolution color photograph of a complete fossil skeleton. His nerves turned to relief as he shared his secret with Hurum. "This fossil needs a good home," he said. "When I gained access to the fossil, I was very excited about it going to scientific research."

The photograph was of Ida, fossilized after her tragic death.

Hurum was shocked. He knew right away that he was looking at a primate, the order of mammals containing humans,

12

because the big toe was standing out from the foot and the fingers had nails instead of claws. Since the fossil appeared so complete and well preserved, he knew that it had probably come from the Messel Pit. The site had produced some of the world's most complete and articulated fossils, but this one was truly remarkable. The unique geology of the pit allowed him to conclude that the fossil was from the Eocene epoch, or the dawn of recent life. If he was right, it could represent a major scientific breakthrough.

The Eocene, which lasted from 55.8 million to 33.9 million years ago, was a crucial turning point in evolution. Though dinosaurs and mammals had coexisted briefly, the world belonged to mammals now. The first prototypes of the creatures that modern man shares the planet with were emerging, notably the primitive primate. Although, because of gaps in the fossil record, paleontologists have had to hypothesize about what happened after the primitive primate, they have determined that by 40 million years ago, there were, as we know, two distinct primate groups: those with wet noses—lemurs and lorises; and those with dry noses—tarsiers and apes and monkeys.

At some point during the Eocene, this important split in primate evolution occurred; without it, humankind as we know it would not exist. Until the fossil in the photograph was found, no complete skeleton had ever been discovered of an "in-between" species to prove this split. Hurum was fast concluding that the specimen he was looking at could be one of the holy grails of science—the "missing link" from the crucial time period.

Hurum marveled as he studied the picture. She was lying on her side, so Hurum could see one foot and the hands. He could see impressions of the rounded fingertips so typical of nail-bearing fingers. Even the tiniest vertebra was visible. He could see the tail clearly, as well as the fur on the body. The jaw was intact, and it appeared that the teeth were too, which he knew was of great scientific importance because teeth hold the greatest clues to an animal's lifestyle. The shape and stage of the teeth can reveal the age and the diet and also verify that a fossil is of a primate. Amazingly, he could also see the stomach content, evidence of the last meal. She almost looked alive.

"This is so beautiful," he said, having tuned out the fair swirling around him. "It's like finding the lost ark."

The find represented a once-in-a-lifetime experience for any paleontologist. Nobody had ever seen anything at all like it before, except for the private collector who owned it, Perner, and now Hurum and his colleague. After Perner showed him pictures of the hands and one of the feet, making it clear that the specimen had the fingernails and opposable toe of a primate, Hurum knew he had to protect the fossil at any cost.

But the asking price was steep—$1 million. The Oslo Natural History Museum had never paid more than $15,000 for a fossil. Hurum asked Perner to give him until after Christmas to talk to some of his contacts to see if he could raise the money—provided the fossil was genuine.

Hurum couldn't sleep for the next two nights. He tossed and turned, hoping the fossil was real and wondering how he was going to bring this discovery to the world. There were strict rules when it came to fossils like this because so much illegal private collecting went on around the world. Hurum would not be able to formally describe this fossil in any scientific work unless it was legally collected and placed in an official museum. This was to make sure that other scientists could also access the specimen in the future. He knew of complete fossilized dinosaurs that would never become part of the scientific record because they had been collected illegally and were now in the hands of private collectors. He didn't want this fossil to fall into that category.

First he would have to raise the money. The Oslo museum was clearly his first choice, but he was worried that, since the University of Oslo was its primary funding source, it wouldn't be able to put up the money. He began to think of other museums with wealthy sponsors that he could call. He would need to hire the most reputable scientists to conduct CT scans and X-rays to prove the fossil's authenticity. Once that was finished, he would need to enlist other experts in Eocene primates and the Messel Pit to describe the specimen. They would all have to work in unison and then present their findings to the scientific community, which was certain to have its skeptics.

Crucially, Hurum would have to make sure that it was a legal specimen, meaning that it was excavated before the Messel Pit became a protected UNESCO World Heritage Site in 1995. Permits to export it from Germany to Norway would also need to be secured. If the fossil was collected after 1995, it was unlikely that it would ever be allowed to leave the country.

But despite the challenges ahead of him, Hurum knew he must see the fossil in person and ultimately relocate it to a public museum for study and observation. The work he had to do seemed insignificant in the face of his strong hunch that he was about to come face-to-face with the most complete primate fossil ever found—maybe even that of a human ancestor.

CHAPTER TWO

IDA'S STORY BEGINS

When Charles Darwin published *On the Origin of Species* 150 years ago, his ideas were so radical that illustrators needed a simple tool to convey them to the general public. They settled on what has become the classic drawing of the "Ascent of Man," which shows monkeys turning into apes, and apes turning into an upright man. Although Darwin believed the drawing wasn't entirely accurate, he accepted that the public needed a tool to grasp this radical idea. But scientists now know that man didn't evolve from other primates, he split with them. These splits occurred throughout the time of life on Earth as each line of creature evolved, and at each split, there is a hypothetical missing link—a creature that is the first step toward a new kind—known as a *transitional species*.

Monkeys, lemurs, chimpanzees, and humans are all primates, and like any family, we must share a common ancestor, but just what the common ancestor is and where it evolved have remained great mysteries. The primate fossil record is so sparse that only around fifty significant specimens exist from the past 5 million years. The most famous is Lucy, the 3.2-million-year-old australopithecine discovered by Donald Johanson in November 1974. Lucy revolutionized science by providing the first evidence of a primate that walked upright—a crucial link in our own evolution that distinguishes us from all other primates. But even Lucy, considered a remarkable specimen, was only 40 percent complete.

Johanson found Lucy in East Africa, widely regarded as the cradle of mankind. Because East Africa has been the site of virtually all humanlike fossil finds, most scientists have hypothesized that that is probably where the human line of primates evolved, and thus most likely the site where earlier primates developed into the distinctive wet- and dry-nosed groups.

Johanson's discovery dramatically furthered our understanding of modern primate development, but early primate development has remained a complex mystery. The wet-nosed primates are the lemurs and lorises, and the dry-nosed primates are the tarsiers, monkeys, apes, and humans. But where and when did the primate family tree split into its two main strands?

Jørn Hurum believed that Ida could not only answer that question but in doing so could become one of the recognizable icons of science in the twenty-first century. He also knew that the announcement of her discovery could be the beginning of a controversy. The battles over new scientific discoveries can be bitter and competitive, but in the best of circumstances, underlying all the recriminations, the noble pursuit of knowledge for the betterment of science is paramount. Ida's story, however, was complicated. Not only was she dug out of the

Messel Pit in Europe and not found in East Africa, but she was found by an anonymous amateur who had kept her hidden from the world for years.

Hurum knew that if this creature that lived 47 million years ago in a tropical rain forest before succumbing to a tragic death was about to birth a revolutionary chapter in our understanding of the human family tree, great scientific care would need to be taken in both her study and her being revealed to the world. Ida's short life, all that she encountered, and her long journey across millions of years to today would have to be meticulously validated, documented, and explained.

"I knew that I had to get a good team together to work on it," Hurum recalled. "This was an opportunity of a lifetime."

Hurum was facing a ticking clock. He needed to complete the purchase of Ida before the fossil dealer Thomas Perner's six-month option with the private collector expired, and he wanted to carry out the transaction in secrecy because he feared someone else would learn of the discovery and snatch it out from under him. Perner was very reliable, but Ida's keeper was a wild card who could have been pursuing any number of agendas. In fact, Hurum told his wife the briefest details about the find, as much out of superstition as paranoia.

After the Hamburg fossil fair, Hurum returned to Oslo and met in mid-December of 2006 with the Natural History Museum's director, Professor Elen Roaldset, to enlist her support for his funding pitch to the University of Oslo. Hurum walked into her office, and the director greeted him by asking if he had found any interesting fossils at the Hamburg fair. Hurum sat down and got right to the point.

"I saw a picture of the most incredible fossil I am ever going to see in my life," he began. "It's for sale, but I don't think we can afford it. This is a specimen that can really change history."

After Hurum told her the entire story, the normally reserved and academic-minded Professor Roaldset got a twinkle in her eye. The Oslo museum was already well respected, but she knew that a discovery like this could make it famous. She later told a friend that she reasoned, "We're not a museum known around the world like the Louvre, but this could be our *Mona Lisa*."

Professor Roaldset asked Hurum if he could make a presentation to the museum's board of directors at their meeting in five days. Hurum immediately agreed. He rushed to the phone and appealed to Perner to send the photographs of the fossil in strictest confidence so he could show them to the board. Perner agreed, and Hurum began preparing a brief history of the Eocene and Messel Pit for the meeting to provide a context for the discovery.

Hurum remained skeptical that he could convince the board to lay out the money. He had not seen the fossil. In fact, he didn't even know its whereabouts. No X-rays or CT scans had been taken. He wasn't sure it was a legally collected specimen or that the German government would permit him to remove it from the country. All he had were a few photographs and his passion. He decided that his case was best made by comparing Ida's scientific potential to the Rosetta stone, the ancient tablet that opened up doors of study in language for decades to come.

Much to his surprise, given the seven-figure asking price,

the pitch worked. The board decided an hour after his presentation that he could begin negotiating to buy the specimen. They insisted that he check its legal status and the export laws and, of course, that he make certain that it was not a fake.

As Hurum put the authentication process in motion in early 2007, he set out to form an international collaboration of authorities to concentrate on Ida's life and remains. He wanted to recruit scientists who would both help verify authenticity for the purchase and become part of a team that would undertake a complete examination of this miraculous discovery.

Hurum first contacted Professor Wighart von Koenigswald in Bonn and asked him for advice. In 1997 he had been the second examiner on Hurum's PhD thesis on early mammals, and he had worked on different aspects of the Messel locale for many years. He suggested Hurum contact Jens Franzen, affectionately known as "Mr. Messel." With a full gray beard and Coke-bottle glasses with slightly tinted lenses, Franzen has the demeanor of a kind grandfather. He had led the protests against converting the Messel Pit into a landfill, and his work had helped stave off the government's action. Franzen had spent much of his life studying the Messel Pit and its treasure trove of fossil findings, and Hurum knew that his institutional knowledge would be invaluable.

Franzen had worked for more than thirty years at the research institute in the Senckenberg Museum in Frankfurt, and he had overseen the Messel Pit's sanctioned excavations from 1975 to 1984. His understanding of the true scientific value of the Messel Pit had begun in 1973 when a private collector showed him the head of a fossilized bird complete with preserved feathers. For the next twenty-seven years, he had fought for the preservation of the pit and studied its history.

Though Franzen had recently retired, when Hurum contacted him, he was thrilled with the opportunity to participate. The day he received Ida's picture and the invitation from Hurum was his seventieth birthday, and he e-mailed Hurum to say, "This is the best birthday present I have ever had!"

It was as if his life's work had come together in one specimen. During those years when he was trying to save the Messel Pit, he had unknowingly been fighting for a fossil that could be the earliest, most complete primate ever found.

"I'm talking about the eighth wonder of the world," Franzen said of the fossil of Ida. "We have not only for the first time a complete skeleton of a primate from that early age, but there's even the complete outline of the soft body up to the tips of the hairs. So we can really talk about what this animal looked like—for instance, how large the ears were and how long the fur was."

To authenticate the bones of the fossil, Hurum needed an X-ray and CT specialist. Franzen suggested Dr. Jörg Habersetzer, a longtime colleague of Franzen's at the Senckenberg Research Institute and a world-renowned expert in the CT scanning of fossils. Habersetzer has the air of an academic, with a neatly trimmed beard flecked with gray, and he speaks with convincing authority about his craft. He too immediately accepted Hurum's invitation.

Hurum knew that a learned and respected primate expert was essential to the project's credibility. Through Wighart von Koenigswald, Hurum was put in touch with Professor

Philip Gingerich, an expert in the field of Eocene primates, who immediately agreed to participate.

A professor of geological sciences and director of the Museum of Paleontology at the University of Michigan, Gingerich had done extensive work on the environments and evolution through the Paleocene-Eocene transition. He had studied the evolution of archaic whales for more than twenty-five years, collecting specimens in Pakistan and Egypt. In a groundbreaking find in 2000, he discovered fossils that confirmed the assertion by molecular biologists that whales evolved not from mesonychids, extinct wolflike animals, but from artiodactyls, the ancestors of hippos and camels.

As a bonus, Gingerich's wife, Dr. Holly Smith, was one of the world's foremost experts in the dentition of primates. A comprehensive study of Ida's teeth would also be critical in the team's findings and conclusion. Because Habersetzer would create a detailed, computerized three-dimensional image, the couple could do most of their work in their labs in America.

The "dream team," as Hurum colloquially dubbed his group, was now set to undertake the great challenge of detailing exactly what Ida was.

In May 2007, Hurum and Professor Roaldset went to Thomas Perner's house in Bad Homburg, a wealthy suburb of Frankfurt, to see the fossil for the first time.

After they arrived, Perner explained that the collector would allow the fossil to be out of his sight for only a short period of time. Perner then got in his car and drove to the collector's house to retrieve the fossil while Hurum and Professor Roaldset sat in Perner's living room and chatted with Perner's wife.

An hour later, Perner returned with a small plywood crate that was shaped like an elongated wine box. Considering the box contained such a precious and expensive artifact, it was of dubious durability. Placing the box on a table, Perner opened it and carefully removed the plate containing the fossil, which was wrapped in a towel. The plate itself was twenty-six inches (67 centimeters) long and sixteen inches (41 centimeters) wide, and it was just over half an inch (2 centimeters) thick.

"Oh, it's so small," Hurum commented.

Hurum immediately saw that the fossil was better preserved than he had imagined from seeing it in the photographs. After studying the fossil plate for a minute, he felt a sense of relief sweep over him. To his trained eye, from his years of experience dealing with ancient fossils, he could tell that it was not a fake. But not only that, the fossil was so complete he could almost stare into its eyes.

Hurum and Professor Roaldset called the University of Oslo accounting department from Perner's home and instructed them to transfer the first payment. Perner then left to return the specimen, and Hurum and Professor Roaldset headed for the Senckenberg Research Institute in Frankfurt to make arrangements for the delivery and scientific examination.

Two days later, Perner delivered Ida to the Senckenberg institute in Frankfurt, where she would spend the summer undergoing tests in complete secrecy. Franzen dated the artificial matrix — to make sure it was collected before 1995 and was therefore a legal specimen — while the X-ray process was done under the direction of Jörg Habersetzer. The work was performed in secrecy to prevent any leaks about Ida's possible identity.

The first order of business was to ensure that the fossil was completely genuine. Even though Hurum was certain the specimen was not a fake, he could not determine with the naked eye whether parts of it had been doctored in any way. He had heard stories of private collectors filling out the bone structure of fossils and then trying to sell them as complete specimens. Since a fossil X-ray is far more detailed than what humans typically receive from their physicians, Habersetzer would rely on the results to tell him if the fossil bones were real.

"When we study these kinds of fossils, we apply three different radiological methods," Habersetzer explained. "The first is that we bring a plate or a film very close to the fossil plate. This is called *contact microradiography,* meaning that the resolution of this type of radiograph is extremely high. The second method is *enlargement radiography.* That means we have [an] X-ray tube with a very, very small focal point, [and then] we can make an enlargement tenfold or twentyfold. That means we can use [the] X-ray apparatus as a kind of microscope. And finally, we can use this extremely high-resolution X-ray tube in combination with a normal CT scan."

Habersetzer would then feed the X-rays and CT scans into a computer. Because the specimen was so thin, computers were used to make an enlarged, three-dimensional model that would allow the team to reconstruct the primate's structure. This complex and intricate process could take several months, and because the glass fiber cuttings used in the polyester to strengthen the plate cause reflections in the X-rays, these fibers had to be digitally removed from each CT scan so that a clear model could be created.

"We can then analyze the data and go deep inside the primate," Habersetzer continued. "We can reconstruct the teeth and place them exactly where they should be in the skull, even though they've now been crushed flat. We can look at the internal bone structure, telling us what kind of stresses [were] endured."

From the initial X-ray process, which took only a month, Habersetzer concluded that the specimen was 100 percent real and that it had been collected from the Messel Pit.

Hurum was ecstatic at the news. He instructed the University of Oslo to wire the final payment to Perner, and he prepared to secure an export license. Hurum was never told the collector's identity, but for the export license, Hurum needed only proof that the specimen was extracted from the Messel Pit prior to 1995. The pit's previous owners had relinquished all rights to any fossils collected during its ownership. The work to prove the fossil's date needed to be conducted by a neutral party. At the time, Hurum and the Oslo museum were working with the Hessisches Landesmuseum Darmstadt, which has one of the largest collections of Messel fossils in the world, to bring the exhibit Messel on Tour to the Oslo Natural History Museum in the summer of 2008. The Darmstadt museum employed several experts, so Hurum enlisted its scientists to determine a collection date for his specimen.

The Darmstadt paleontologists were quickly able to date Ida's extraction from the Messel Pit to around 1982 because of the preservation technique used by the collector. At that time, an artificial resin was frequently used to embed specimens and to keep them from drying out after they were excavated, and glass fibers were then placed in the resin. The artificial matrix of Ida had both. No other fossils in the world are preserved this way.

Those findings were reported to the Hessen government,

and Hurum's export request moved smoothly through the bureaucracy. Because Ida was taken from the pit prior to both the Hessen government's acquisition of the pit and its being named a protected site, the specimen had been legally collected and therefore could be purchased and exported. The Hessen government issued the license. Amazingly, it was free and there was no export tax.

Ida was now going to tell her own story. The CT scans taken by Habersetzer were made available to the dream team for examination in their own labs. Each scientist would do his or her own analysis, and periodically they would meet—first at the Senckenberg Research Institute in Frankfurt and later at the Oslo Natural History Museum—to pool their results.

Habersetzer continued work on the three-dimensional computerized model of the specimen that would allow the team to reconstruct the primate in its entirety and study her anatomy. Franzen focused on her life in the Messel Pit, and von Koenigswald made the complete study of what happened after Ida's death, known as *biostratinomy*. Philip Gingerich worked on trying to place Ida on the evolutionary tree, and Holly Smith examined her teeth to verify her age. Gingerich and von Koenigswald were familiar with each other's work process because they were also deep into preparing a scientific paper on pregnant fossil whales. "Ida is really key to understanding all of the Messel primates and by extension, all of the Eocene primates," Gingerich declared.

Ida's basic details drove the research. The X-ray confirmed that she had no bacculum, or penis bone, which would have been fossilized had it ever been present. From end to end she measured almost twenty-three inches (58 centimeters). Her body was roughly nine inches (24 centimeters) long. Comparing the length of her legs with the length of her arms provided the team with an intermembral index, a process used for all known animals to determine their locomotion. Ida's legs were longer than her arms, which told the team that she was a leaper and a clinger. Further evidence came from her long fingers and toes and opposable thumbs; these allowed her to grip and stand in the trees. Her long tail would have helped with balance and steering during leaping.

"Looking at the skeleton, we can be sure that it was living on trees, because when you are looking at the thumb and also at the big toes of the feet, you can see that these were grasping hands and grasping feet," Franzen explained. "So these were feet constructed for an animal living on trees. No doubt about that."

A noticeable physical oddity was that Ida's left leg was broken just below the knee. After studying the point of the break, the team came to the conclusion that it wasn't from a hostile action such as a crocodile bite. The break had been such a clean snap that they concluded it occurred while her remains were being pulled from the pit's shale floor.

Since the preservation and taphonomy, or study of the processes that affected her as she became fossilized, verified that Ida came from the Messel Pit, Franzen was immediately able to determine her age at 47 million years old. "We know the exact age of the Messel Pit because in the year 2001 we drilled a borehole in the center of the site and extracted volcanic rocks that were ejected during the eruption that formed the lake," Franzen said. "Isotope composition of crystal in the rocks dated it at approximately 47.2 million years old. That was when the crater was formed, and we know that life there existed for about a million years."

The conditions that created the oil-rich rocks, which at one time provided some of Germany's oil supply, were also perfect for preserving long-lost life. Because of the lake's lack of oxygen at the lower levels, the creatures that died and sank to the muddy bottom remained untouched. When the lake eventually filled up, the mud and algae were compressed into shale. Plant and animal remains were preserved in dramatically intact form, which accounted for Ida's being a nearly complete specimen. Perhaps even more amazing, many of the pit's specimens survived continental shifting and the splitting up and shattering that can result so easily from tectonic plate movement.

"The fossils that are found in the Messel Pit have unrivaled preservation," Franzen continued. "They are often almost complete, and they even have their soft body tissue preserved, sometimes even their gut content." Fortunately, the gut content appeared to be present in Ida.

For such a rare find, Franzen actually saw something familiar about these particular stomach contents. In 2000 a specimen had come to Franzen's attention. It had been bought by Dr. Burkhard Pohl at the Wyoming Dinosaur Center from a private collector in Frankfurt. The specimen was less than complete—in fact, large sections had been reconstructed in polyester instead of bone. However, the gut contents were complete. Little did Franzen know that there was another full half to this specimen, a half that wouldn't come to light for another three years—Ida. And this first specimen became known as slab B.

When Franzen examined the gut contents on slab B in 2003, he thought a fish scale was attached to the specimen. However, upon closer examination through a high-power microscope, he saw that the inner structure was not that of a fish scale. "They were cell structures telling me that this must be the rest of a plant or, more precisely, a fruit." Because other specimens from Messel had shown insects preserved in the gut contents but this one did not, Franzen was able to conclude that this creature probably ate only fruit and plants. With this knowledge from slab B he could immediately shed light on Ida's lifestyle with the team.

Dr. Smith, the team's dentition expert, was cross-checking the idea that Ida was a fruit eater by giving her a scientific dental examination. "Many of the dental patterns that are present in modern primates have been observed throughout the fossil record, but not many fossils are this complete," she said of Ida.

Teeth hold some of the biggest clues about an animal's lifestyle. Different branches of the primate family tree adhere to different tooth patterns, and the teeth are shaped according to the role they played in the animal's diet. Tooth enamel is the toughest material in a mammal. Because it is 98 percent apatite, a mineral, it is practically rock even before fossilization begins. As a result, teeth preserve very well, even over millions of years. Ida is also a nearly complete fossil, so the team's dental evidence can benefit from other skeletal clues.

The shape of a creature's teeth is directly related to the food it eats. Primates with pointy teeth tend to eat insects, the sharp edges slicing leaves and food. Rounded teeth denote fruit eaters. Smith concluded this fossil was predominantly a fruit-and-leaf eater, which was confirmed by Franzen's study of the gut content.

"It's not unusual for fruit eaters to have a mainly nocturnal lifestyle, which other features [in Ida] support," Smith added. "The fossil specimen has very large eyes for its skull size, which leads me to suggest that she was probably nocturnal."

The X-ray of the teeth revealed that Ida was a juvenile. She still had several milk—or baby—teeth in her jaw, but she had developed her first and second molars. However, the third molar was still in the bone, so the question arose of what the eruption patterns of the teeth told the scientists about her age at the time of death.

"We can tell from its teeth that's it's a juvenile," Gingerich said.

"But how old?" Hurum wondered.

Franzen then came up with a practical explanation. "Do you remember when your child lost their teeth?" he asked the two. "There was a few months' gap between the first teeth falling out and second ones coming through, so the child had a mix of baby and adult teeth."

After further examination and discussion with Smith, that is exactly what the team concluded about Ida. She had a mix of adult and baby teeth, and she was on the cusp of losing the last of her baby teeth. A few more months of life and her molars would have erupted. And, as it turned out, if she'd been a few months older, she might never have become the specimen we have today. It was at this point that the name Ida popped into Hurum's head. He was looking at a little girl, and he had his own little girl at home named Ida. Like the fossil, his five-year-old daughter had a mix of primary and permanent teeth. She and Ida were at the same stage developmentally.

Hurum had been calling the fossil "Little Miss Messy"

after the Messel Pit, but he wanted to settle on a populist name that would immediately communicate to people that she was a real creature, much the way Lucy had done. Hurum kept returning to the classic children's book *Pippi Longstocking* by Swedish author Astrid Lindgren because the three-dimensional image of his fossil looked like Pippi's monkey. He had considered naming the fossil Nelson after the monkey, but the surname is so common in Sweden that several other species already carried the name.

Aside from the personal overtones, the name Ida satisfied his other concerns too. It is a Germanic name, and Hurum felt it was important to pay tribute to the fossil's home country.

Before making a final decision, Hurum discussed privacy concerns with his wife, Merethe. They both concluded that any initial media fascination with their own daughter would be short-lived and the fossil itself would take center stage. Hurum then broke the news to his daughter and showed her a picture.

"It was just a skeleton to her, but she did tell her kindergarten class that a dead monkey was going to be named after her," Hurum says.

That sealed it. From that moment forward he always called the fossil Ida.

For something whose existence on Earth spans 47 million years, Ida had her busiest few months in the summer of 2007. By September, all the X-rays and CT scans were completed and the three-dimensional computer model of the skull was being built, so it was time to move Ida to the Natural History Museum in Oslo. For Ida, this ultimately would be an ideal permanent home, where scientists and interested viewers could visit her.

The paleontological section of the Natural History Museum is both a serious place of scientific study of the world's origins and a fun house for anyone curious about the wonders of history. It is located inside the museum's storied botanic garden, which contains eighteen hundred different plants arranged according to genus and species. Visitors are greeted by the skeleton of a 67-million-year-old *Tyrannosaurus rex*. They move on to encounter a 400-million-year-old sea scorpion, a 1-million-year-old giant sloth from South America, and a dinosaur nest containing six eggs, as well as fossils of plants, mammals, fish, amphibians, and humans. Ida, however, would be the star attraction.

Ida was the most precious of cargo, so great care had to be taken to transport her from Frankfurt to Oslo. A company that specialized in moving precious artifacts was contracted. The plywood crate that Perner had used to transport Ida in his car was scrapped, and a custom wooden crate with firm inner walls was built. The plate containing the artificial matrix was then covered with Bubble Wrap to absorb any vibration on the 680-mile (almost 1100-kilometer) flight.

Nine months after Hurum had first seen pictures of the fossil, it arrived at the Oslo museum. The crate was unloaded and placed in Hurum's office, a comfortable work space with mahogany walls and glass-encased bookshelves. He invited Professor Elen Roaldset and several of the museum's staff for the private unveiling.

Hurum opened the crate, carefully lifted the fossil plate out, and placed it on a dark wood table in the center of the room. He then popped open a bottle of champagne and offered a toast to Ida.

Glass in hand, Hurum declared, "What we are looking at is the most complete skeleton of something that is in our own lineage."

But on the issue of what that "something that is in our own lineage" was, or where this tiny primate fit into the evolutionary tree, the scientists soon hit a stumbling block in their research. The initial CT scans suggested that they were looking at a lemur's predecessor. Clearly, they would need to do further comparisons to answer the big question: Was Ida a human ancestor?

The team first established that the size, lifestyle, and diet of Ida were comparable to the Eastern woolly lemur of Madagascar. Then their next challenge was to investigate the fossil bone-by-bone and compare its anatomy to the anatomy of all known primates.

There are two very distinctive traits of a lemur. First, lemurs have a toothcomb, a set of long, flat, forward-angled teeth in the lower jaw that are present at birth. Ida doesn't possess these. The second characteristic is a grooming claw, an elongated claw on one toe that lemurs use for cleaning fur that cannot be reached by the toothcomb or after defecating. If it were present on the fossil, it would be clearly visible, since on all the other digits nails can be seen. Ida doesn't have a grooming claw.

Gathered in Hurum's lab in the Oslo museum, Hurum and Franzen reexamined the X-rays for further clues about Ida's life and death. There was a bump on Ida's right arm that they had originally concluded was a siderite concretion affixed to the bone, siderite having been formed when the carbon dioxide reacted with the iron in the lake. However, a closer examination showed that Ida's right wrist was fused from a break and had not fully healed. The lower bone of the left arm was also slightly displaced, meaning that it had been broken at one time and had partly healed.

"She was not very mobile," Hurum said. "She couldn't climb or grasp with her right hand, which caused her death."

Franzen agreed that her death was an accident. "[She] came down from the tree and was drinking at a small creek running into Lake Messel or even on the shore of Lake Messel itself, and perhaps because [she] was just bringing [her] head down to the water to drink, [she] may have gotten into the poisonous layer of gas and lost consciousness."

As the beginnings of a narrative of Ida's life and death came together, Habersetzer's earlier work on Messel bat specimens answered another critical question. Habersetzer had found that the lake was covered by a low level of poisonous carbon dioxide gas. Many animals avoided the gas by moving out of the toxic areas into the trees. But an animal with a broken wrist and a weak arm would not have been able to climb quickly enough to escape.

Gingerich put the team's work in the broader spectrum of science: "This specimen is helping us understand how primates developed [and] how the most successful animal to have ever lived on the planet started."

The study of Ida continued at the Oslo museum throughout the fall of 2007. The fossil was photographed from every conceivable angle to map its scientific description. A silicone mold was made to allow all of the scientists on the team access to her in their respective offices.

The dream team was also beginning to assemble its findings and write a scientific paper, which would provide context by giving readers an overview of life in the Messel Pit during the Eocene and then give details via the bone-by-bone analysis. As they set down their conclusions for the world's scientific record, they knew that their work had produced an incontrovertible result and a scientific first: Ida was, in fact, the offspring of primates that lived 47 million years ago.

"It's the oldest, most complete primate ever, so really, we're making history," Hurum said of the dream team and their work.

Hurum and each of the members of his team had been excited by the find from the start; Ida's age, her condition, and her discovery in Messel were all tremendously significant scientifically. But as they drafted the article that would announce her, committing the discovery to paper for the world, the brain trust began to consider the broad potential impact of their work. "When we publish our results," Franzen said, "it will be like an asteroid hitting the Earth."

For all of the vast knowledge of the Eocene, the Messel Pit, and primate evolution that went into the study of Ida, the scientific paper's early draft begins in the simplest language: "The first nearly complete skeleton of a juvenile female prosimian is described from the early Middle Eocene of the Messel Pit (Germany). Ida was living as a vertical clinger and leaper on a middle floor of the Messel rain forest. The specimen was discovered, recovered, and prepared by private collectors. It comprises gut contents as well as a complete soft body outline."

But in those few words lies a complex, dramatic story with great implications for how we view evolution. In order to put in context a detailed look at Ida's highly unusual, specialized anatomy, her place in the primate family, and even her last meal, we must start in the Eocene, the era in which she lived and the one that so dramatically shaped the development of primates and indeed of the mammals that we recognize to this day.

CHAPTER THREE

IDA'S EOCENE WORLD

Though it was tens of millions of years ago, the Eocene had some modern aspects. Most of the plants would be roughly familiar—even if many of them were growing where they wouldn't grow today. In general the birds were modern, though they included some that are nothing like anything that lives today, and ones that do belong to modern groups—like the Eocene woodpeckers and parrots—were nothing like their present-day descendants. The mammals included some that were left over from dinosaur times and were doomed to go extinct within the Eocene or shortly after; some—including some huge hoofed animals—that arose afresh only in the Paleocene but were also doomed to extinction; and some that were the ancestors of modern types, including some very recognizable primates. But if we could go back easily and stand on the Eocene Earth, we would notice one thing first: the Eocene was hot.

In fact, the first 5 million years of the Eocene—the time that included Ida's brief life—saw the warmest spell in the whole of the Cenozoic era. Palm trees and other tropical plants put in a brief appearance in Alaska and Siberia. Palms looked out from the northern coast of Canada over an Arctic sea that was totally innocent of ice all the way to Asia. At the other end of the world, the ice melted from Antarctica, and the oceans rose and invaded the land, creating new seas and dividing great landmasses into island continents. Most of the trees and other plants rode out the hot spell more or less unchanged: they simply migrated north or south to find the latitude that suited them best. But the animals, and particularly the mammals, changed absolutely; they were evolving rapidly anyway, but the extraordinary climate accelerated the process. The ancient animals of the Paleocene went extinct during the Eocene with no descendants, but they were replaced by the ancestors of our modern fauna. The new animals flourished, and none more so than the primates. And since Ida and her kind were among the beneficiaries of this climatic influence on evolution, so these many years later are we.

The events that kicked off the Eocene and made it so profoundly different from all that had gone before mimic those of our current global warming crisis. The mechanisms are the same. The only difference is that the events of 56 million years ago were triggered by geological phenomena that were well beyond the intelligence of the creatures of the day, so they just had to go along with them. Though the modern world is too crowded and obstructed with cities and infrastructure for us to look forward to such vast changes, the changes of the Eocene brought benefit to our ancestors.

So, what did happen during the Eocene, and why? We should start 56 million years ago, at the beginning.

What Was the Eocene and
Why Was It So Warm?

The great Scottish geologist Sir Charles Lyell (1797–1875) first defined the Tertiary in 1833 as the period that came after the demise of the dinosaurs; the rocks from the early Tertiary were conspicuously devoid of dinosaur fossils. He divided it into four epochs: the Eocene, Miocene, Pliocene, and Pleistocene. The term *Eocene* comes from the Greek *eos,* meaning "dawn," and *kainos,* meaning "recent," so Eocene meant "first new life"—the first wave of creatures after the great age of reptiles. Throughout the nineteenth century, the Eocene was further divided into the earlier Paleocene and later Oligocene. The dating methods of the nineteenth century were very rudimentary, and the dates were put in later. We now understand that the Paleocene ran from the time of the K-T boundary (which is what we call the transition from the Mesozoic to the Cenozoic era, from the Cretaceous to the Tertiary period; the *K* comes from the German word for Cretaceous) 65 million years ago, until around 56 million years ago; the Eocene stretched from 56 million to around 34 million years ago; and the Oligocene emerged as a long and comparatively dreary interval from 34 million to around 26 million years ago. At 47 million years old, Ida lived in the early Middle Eocene.

The Cenozoic era began with a bang, literally—one of the biggest bangs the world has ever experienced. Theory has it that the world was struck by an asteroid—a very big one, though actually quite small relative to the tremendous damage it did. The asteroid struck what is now the Yucatán Peninsula of Mexico, sending so much debris into the atmosphere that it changed the climate more or less instantly from warm to cold. The immediate impact for life on Earth, although the precise reasons are still debated, was the instant extinction of the dinosaurs and the other big reptiles, including the plesiosaurs and mosasaurs, which lived in the oceans.

During the Paleocene, the epoch that followed this putative disaster, things slowly returned to normal, which means that the world warmed up again. We know from the fossil record that the world has sometimes grown hotter and sometimes colder during the past few hundreds of millions of years, and cores of ice taken from Antarctica and Greenland, together with other, more recondite evidence, show that the hot periods have always correlated with high concentrations of CO_2 in the atmosphere, and cold periods with low concentrations of CO_2. Levels of CO_2 can fluctuate for various reasons, but whatever the cause, the greenhouse theory certainly seems to apply. It appears that CO_2 increased throughout the Paleocene, and global temperatures rose with it, and volcanoes likely gave a final boost at the end.

But CO_2 is not the only greenhouse gas. Together with water vapor, it is the commonest and at present the most important of the warming gases, but it is by no means the most potent. More effective by far—ten or twenty times as effective—is methane, CH_4. The world harbors vast stores of CH_4, and every now and again when conditions are right, those stores can be released—and if that happens on a big enough scale, then the whole world is cooked, in no time at all. It seems that this is what happened at the start of the Eocene, and the consequent burst of warming is what distinguishes the Eocene from the balmy but far less dramatic Paleocene.

Methane forms when organic material is left to decay under anaerobic conditions. It forms in vast quantities—many millions of tons—when planktonic diatoms (which have shells made out of silica) and other such creatures sink to the bottom of deep, still oceans, where there is no turbulence to bring oxygen-rich waters down from the surface. If conditions are cold enough, the newly formed methane becomes trapped in the permafrost that forms in the mud of ocean beds and becomes what is commonly known as *methane ice*.

This methane ice can last indefinitely, but if the ocean warms significantly, the methane ice beneath the ocean bed melts. Then the methane, many millions of tons of it, comes free, drifts to the surface, and enters the atmosphere. Because methane is such a potent greenhouse gas, the world warms up, quickly and dramatically. This is what happened at the end of the Paleocene. Fifty-six million years ago, the general temperature was high enough to melt the clathrate—the icy, methane-hosting compound at the bottom of the sea—and the methane burst forth.

All this can be inferred from an eminently subtle piece of geochemistry. All chemical elements exist in more than one form, and each form is known as an *isotope*. Different isotopes of any one element are effectively identical in their chemistry, but they differ in atomic weight. The commonest form of carbon in the world has an atomic weight of 14, and is known as carbon 14. But there are two lighter forms—carbon 13 and carbon 12. Methane that has been trapped has a higher proportion of carbon 12 than normal, and geochemists have found that rocks from 50 million years ago contain more than the expected amount of such carbon 12. The most likely explanation is that the atmosphere suddenly received a burst of methane gas. Since the fossils tell us that the world warmed up at this time—tropical forms suddenly started to appear at high latitudes—the facts fit together very nicely.

There were two bursts of methane "out-gassing" from the ocean at the start of the Eocene, each lasting about one thousand years and separated by about twenty thousand years. As a result, the temperature of the sea rose 6 to 8 degrees C. This was enough to cause dramatic changes in the ocean currents, which in turn changed the way heat was distributed around the world. As the sea warmed and the heat became distributed more evenly, it became less turbulent. The warm water sat on the top, and the water at the bottom, no longer stirred from above, was virtually devoid of oxygen. We can see the results in the foraminifera—the single-celled marine organisms whose fossil skeletons formed the world's chalk. The planktonic foraminifera, floating near the surface, evidently flourished in the warm spell. But between a third and a half of the benthic kinds—the species that lived in the depths—died off. Indeed, more benthic foraminifera died during the global warming that began the Eocene than in the K-T catastrophe that saw the end of the dinosaurs.

Though the fate of ancient deep-sea single-celled organisms emphasizes just how dramatic such events can be, it's the developments on land that concern us most. The tropics stayed much as they were during the Eocene, and today they are much as they were then: a continuous belt of rain forest all around the globe. The poles were affected most significantly because of feedback effects. Ice reflects solar energy, and because of this, the poles are slow to warm. But as the ice melts because of rising CO_2, the poles warm quickly and dramatically. By comparison, the climate of the Eocene world as a whole was more homogenous than at any time since—temperate right up to the poles and positively sub-tropical at times up into the Arctic Circle. Being warmer, the climate was also wetter, since there is more evaporation from the sea when the temperature goes up. These changes are reflected dramatically in the period's living creatures.

The Life of the Eocene

The plants that shared the planet with Ida and the early mam-mals were not too affected by the dramatic global warming kicking off the Eocene. There were plenty of ancient non-flowering plants such as conifers, cycads, and ferns, just as there are now. The angiosperms, the flowering plants, had already come into their own in the Cretaceous (in late dino-saur times), and in general they sailed on merrily through the Paleocene and Eocene. Most of the ones that were around then would be very recognizable today. On the whole, the plants, and especially the trees, enjoyed the warmth and wet, and they spread themselves out, farther and farther toward the poles. There must have been forest everywhere there was land. These forests would have been mixed, with oaks and chestnuts sometimes growing right alongside palms and trop-ical laurels.

The animals of the Eocene showed mixed fortunes. In the mid-nineteenth century, zoologists realized that dinosaurs disappeared suddenly from the fossil record, to be replaced almost immediately by a growing and increasingly varied succession of mammals that, as their fossil skulls suggested, tended to get brainier and brainier as the millennia passed. The Victorians did not know how long ago this replacement began to take place, but it now seems to have been about 65 million years ago.

Why had the dinosaurs been replaced by mammals? In his *Origin of Species,* published in 1859, Charles Darwin told us that all living creatures are bound to compete, one with another, and that only the best-adapted types survive. The philosopher Herbert Spencer in the 1860s summarized Darwin's idea of natural selection as "survival of the fittest," an expression that Darwin then adopted. In Victorian times, *fittest* commonly meant "suitable" or "apt." But it also, of course, has connotations of health and raw strength. Darwin expressly denied that natural selection necessarily produces creatures that are notably superior to their predecessors. He had spent a long time studying barnacles, after all, and barnacles are astonishingly successful—we find them everywhere. But they descended from free-living, shrimplike ancestors and became barnacles by losing their brains and sticking themselves head-first to rocks—hardly a great leap forward, but it worked.

With natural selection reduced to survival of the fittest, the assumption was that the animals that came later must have been superior to the earlier ones they replaced. And since in this case the later ones were mammals, and the dinosaurs they replaced were apparently lumbering beasts with tiny brains, the reasons for the takeover seemed all too obvious. Even into and beyond the late 1950s, children were taught that the newly emerging mammals of the late Cretaceous must have ousted the dinosaurs simply by outsmarting them—perhaps by stealing their eggs.

The facts are now interpreted quite differently. In truth, the dinosaurs were by far the most successful big land animals that have ever lived. They first appeared about 220 million years ago and diversified to form many hundreds (and probably many thousands) of species that lived throughout the world. They dominated all the land (and the comparably impressive plesiosaurs and ichthyosaurs and pterosaurs dominated the seas and skies) for more than 130 million years. Indeed, they live on in the form of birds, which are the direct descendants of dinosaurs. It's obvious now too that many of the dinosaurs lived complex social lives, and many were very good parents, just as modern birds are. Fossils of dinosaurs guarding their eggs, replete with fossil embryos, have been discovered. The mammals of the day would not have found it at all easy to steal them.

Now it is clear that the mammals also evolved more than 200 million years ago. In fact, the oldest known mammals could be older than the first dinosaurs. If the mammals were innately so superior, why didn't they, rather than the dinosaurs, take over the world? In fact, for all the time that the dinosaurs reigned, the mammals lived as hole-in-the-corner creatures, none of them much bigger than a badger. We know of no Cretaceous primates, but they probably did begin in late dinosaur times.

In fact, the mammals did not replace the dinosaurs by outcompeting them. Instead, as was first suggested in 1980, the dinosaurs and the other megareptiles were all wiped out by an asteroid, as outlined above. There was little or no head-to-head competition between the dinosaurs and the mammals—or if there was, the dinosaurs definitely came out on top. It wasn't until the stage had been cleared that mammals stepped into the breach.

The first mammals of the Paleocene were still small, but before the epoch was over they had produced some giants. There were no big carnivores among them, but the herbivorous *Uintatherium* and its relatives were the same shape as a modern rhino and even bigger, standing foursquare on pillarlike legs and over six feet (almost 2 meters) at the shoulder.

Such creatures persisted into the Eocene, and big carnivorous animals first appeared then too. They were not modern carnivores but belonged to two orders that both disappeared as the Eocene ended, or not long after. One of them was the Creodonta, one family of which, the Hyaenodonts, were by far the dominant large carnivores of the age. The other group was the mesonychids, which included the biggest land carnivore that ever lived—the spectacular, grotesque *Andrewsarchus,* with a huge crocodile-like head and a vast hyena-like body, culminating in a big heavy tail. The jaws and teeth of the mesonychids clearly show they were carnivorous, but instead of claws, bizarrely, they had small hooves.

Also prominent throughout the Eocene—though much less flashy than the giant rhinos and carnivores—were the multituberculates. These were small, rodentlike creatures with gnawing incisors. Based on the shape of their pelvis, they gave birth to very small and immature offspring, like a modern marsupial. The multituberculates weren't just another mammal; they represent a whole division of mammals quite separate from the modern marsupials and placentals, though not quite as different as today's monotremes (the duck-billed platypuses and the echidnas). They were also, some say, the most successful mammals that have ever lived, common and diverse from the mid-Jurassic (roughly the time of *Archae-opteryx*) right through to the Oligocene, 33.9 million years before the present. But they too are gone.

All these mammals continued to hang around in the Eocene but none of them lived on, or not by very much. Apparently they could not survive the ever-changing conditions—or at least, they did not adapt as well as the new suite of mammals that evolved and diversified as the old-timers faded. By the end of the Eocene, these inheritors included representatives from most of the modern orders: recognizable horses, rhinos, tapirs, proboscideans (the ancestors of modern elephants), artiodactyls (ancestors and relatives of modern cattle and antelopes), and modern carnivores, while rodents increasingly replaced the multituberculates. The whales came on too in a sea freed of ichthyosaurs and plesiosaurs. The Eocene whale, sometimes called *Zeuglodon,* though also known as *Basilosaurus* (though *saurus,* meaning "lizard," is a poor name for a mammal), was up to sixty-five and a half feet (20 meters) long—a respectable size even by modern standards. The Eocene, warm and forested, was the golden age of primates. Ida and her kind truly came into their own.

Because of this eruption of mammalian life, and because mammals are the earth's biggest surviving land animals, and because human beings are mammals, and because mammals dominate so many land environments, the Tertiary as a whole is often called the Age of Mammals. Other beasts have flourished just as impressively these past 65 million years: modern fish, snakes, modern frogs, butterflies, and honeybees—and the birds have been very big players indeed.

It's clear in general that most of the modern major groups of birds were already around by the Eocene. In the modern world, eagles in particular play a huge role in the lives of primates, particularly the arboreal kinds, which must feel reasonably safe from the ground-living predatory mammals but which get zapped in huge numbers from above. One of today's eagles from South America is specifically called the Monkey-eating Eagle. There are no signs of large birds of prey so far at the pit at Messel, but it would be very surprising if they had not been around.

Indeed, at the start of the Tertiary, before the big carnivorous mammals came into their stride, several very large birds almost managed to usurp the megapredator niche. Best known and most convincing of these was *Diatryma,* a huge bird that lived in North America, as well as in France and Germany, making it at least a near neighbor of Ida. *Diatryma* was flightless, like a modern emu—it was much too big to fly—but it was far more formidable. It stood over six feet (up to 2 meters) tall and, unlike an emu, was seriously sturdy. Most impressive, however, was its enormous head—about the size of a modern pony's—with a commensurately huge hooked beak like a giant eagle's. *Diatryma* seems to have been related to modern fowl such as chickens, which might not seem too impressive until we consider the sheer belligerence of chickens in the form of fighting cocks. It is possible too that *Diatryma* used its huge beak as a scythe for slicing coarse vegetation. But the most obvious guess is that it was a predator, reminding us that birds are descended from dinosaurs. Here, perhaps, was a Tertiary reincarnation of *Tyrannosaurus rex*—somewhat miniaturized but formidable nonetheless. Perhaps the nearest equivalent in the modern world is the cassowary, which can disembowel the unwary wanderer with one stroke of its prodigious clawed feet. *Dia-*

tryma, though, probably attacked with its beak. *Diatryma* might have disappeared by Ida's day—no fossils of it are known beyond the early Eocene—but it certainly coexisted with Ida's immediate ancestors.

The Denizens of Ancient Messel

We mammals may like to think that the Tertiary was the Age of Mammals, but the mammals these past 65 million years have certainly not had it all their own way. And a fair slice of the era's diversity is reflected in the fossil record at Messel. The astonishing Messel Pit site is one of the greatest Eocene sites in the world, with truly wondrous fossils, including entirely intact, fully articulated mammals. Ida herself is the only nearly complete fossil primate ever found, from any period; and to find her with soft parts too is a gift indeed. In general, the intact beasts are in a relaxed position, which is typical of creatures that have drowned—especially when they have been anesthetized first.

Until now, the most famous of the Messel fossils were seventy-plus skeletons of ancient horses, including stallions, mares—some of them pregnant—and foals. Also present at the Messel site, though much rarer, are the tapirs, of the species *Hallensia matthesi.* And plenty of bats are also interred at Messel, most looking very modern, and some of them so well preserved that zoologists have been able to study the anatomy of the inner ear that enabled them to hunt insects on the wing at night, by echolocation, as they presumably did.

There are surprisingly few primates among the Messel fossils—given that the forests around the lake would surely have been their ideal territory—and, in fact, only fragments have been found. And this makes Ida even more extraordinary. Indeed, mammals make up only 2 percent of the total fauna found at Messel. There are more than a hundred birds and another hundred reptiles and amphibians—snakes, crocodilians, turtles, lizards, frogs, and salamanders. Again, their fossil remains are largely intact. More than ten thousand fishes have been found. There are also some spiders and a great many insects. Few of the insects are the aquatic types we might expect. Most are fliers, and presumably, like the birds and bats, they were anesthetized by the CO_2 as they flew over the lake, and simply fell out of the air.

Most of the water-dwelling insects that have been found were contained within coprolites—fossil feces—from fish. Careers have been made by examining fossil fish poop for the remains of ancient insects, and many of these wondrous remains can now be seen in Germany, at the Senckenberg Museum in Frankfurt and at the Hessisches Landesmuseum in Darmstadt. The pit, all in all, is a veritable *Konservat-Lagerstätte*—a German term widely adopted in paleontology that means literally "conservation warehouse."

Archaeopteryx, the world's first-known bird, was also found in Germany—at Solnhofen, not too far from Messel. Like Ida, the first *Archaeopteryx* apparently drowned, in this case in a lagoon filled with brine inside a large reef. It looks like a similar scenario. But in truth, the circumstances were very different.

The difference relates to continental drift. As the German geologist Alfred Wegener first suggested in 1912 and as is now universally accepted and well known, the great landmasses have been shuffling around the surface of the globe ever since land existed. Very few people believed Wegener during his lifetime because he could not offer an explanation for how masses of land could possibly move, but now people do believe what he said because we know how it happens. The surface of the Earth is divided into tectonic plates, which rest on semimolten rock known as *magma*, which is constantly churning through the heat of convection, like water in a kettle. The continents, made of rock that is lighter than the magma, float on the top, like froth on a river, and are shuffled around as the magma swirls beneath. The movement is slow, but over tens of millions of years, the distances covered are impressive. In the days of *Archaeopteryx*, 140 million years ago, the world looked very different from the way it does today. The place where Solnhofen is located and where *Archaeopteryx* once lived was the tropics—virtually straddling the equator. That's why it was so hot.

But by Ida's day, a mere 47 million years or so ago, the map of the world looked very much as it does now. There were, to be sure, some important differences. South America, still drifting north after it split from Antarctica, had not yet made contact with North America, so the Atlantic and the Pacific could still flow through the strait that is now bridged by the Isthmus of Panama. The Drake Passage—the wide sea-lane that now divides the southern tip of South America from the northernmost islands of Antarctica—was closed, causing water from the warmer latitudes to the north to flow farther south, which helped to ensure that Antarctica was a great deal warmer than it is now. The place we now call Germany was roughly where it is today. Messel was about 10 degrees south of its current location, about twenty-two miles (35 kilometers) southeast of Frankfurt. But the world as a whole was warm from pole to pole, and at that latitude the

51

climate was subtropical. So Ida and *Archaeopteryx* enjoyed the same conditions.

But as the Eocene drew to a close, the world grew cooler again, and this tropical abundance came to an end, with the lush Eocene landscape retreating back to the equator. Although there have been many warm spells since then, called *interglacials,* the world in general has been growing steadily cooler ever since, culminating in the twenty or so Ice Ages of the past 2 million years. How come?

Why Did the World Grow Cooler?

The reasons that the world began to cool again toward the end of the Eocene, and in general has gone on cooling, are highly complex. They involve subtle changes in the layout of the continents, which have led to shifts in ocean currents; the reestablishment of ice fields, which have led to increased albedo—the extent to which the planet reflects the light of the sun—and have exacerbated the cooling; shifts in the Earth's orbit—sometimes we are farther from the sun than at other times; and so on. But there are two leading theories that are both highly intriguing and also indicators of how very complicated—and surprising—these changes can be. The first involves a tiny floating fern.

The aquatic fern *Azolla* is tremendously important in the modern world. Like most floating plants of all kinds, *Azolla* has air-filled spaces between the cells of its leaves, which increase its buoyancy and allow it to float and maintain its pleasant position in the sun. These damp, air-filled sunny-but-not-too-sunny spaces within the leaf provide ideal conditions for cyanobacteria to grow. Cyanobacteria are the organisms that we formerly called *blue-green algae*. In truth, they are not algae but bacteria, but like algae—and indeed like *Azolla* itself—cyanobacteria contain chlorophyll and they photosynthesize. The particular cyanobacterium that grows in the leaves of *Azolla* is called *Anabaena*. The two organisms have a very cozy setup: a photosynthesizing bacterium is happily lodged within the fabric of a floating photosynthesizing fern.

But there is more to it than that. *Anabaena* does not merely photosynthesize. It also "fixes" nitrogen, which means that it grabs nitrogen gas from the atmosphere and turns it into molecules (or bits of molecules) such as ammonium. This is an absolutely vital contribution to the world's economy. Nitrogen in the form of a gas is useless as far as plants and other living creatures are concerned. But once nitrogen has been locked into ammonium or something similar, plants can use it as a nutrient. This is the source of the nitrogen that they need to make proteins, and nothing is more important to an organism than its proteins. So the *Anabaena* within the *Azolla* leaves not only help the *Azolla* to photosynthesize but also act as the *Azolla*'s built-in fertilizer factory. They provide for free the very element, in the appropriate form, for which farmers pay a fortune to feed their crops.

Because of this symbiotic relationship, floating beds of *Azolla* grow tremendously quickly. In the right climate, they double their weight in two to three days. If they were to grow at such a rate for one hundred days, if the planet could support them, they would weigh more than the Earth itself. Today, *Azolla* with its cargo of *Anabaena* grows in paddy fields. Surplus nitrogenous compounds from the *Anabaena* leak into the surrounding waters and feed the rice—for free. When the ferns and their onboard cyanobacteria die, they rot, which provides more fertilizer. Rice, for example, really can be grown organically, without any added artificial fertilizer (which does not stop industrial growers from adding fertilizer anyway and overriding the natural system).

In the Eocene, *Azolla* grew all over the Arctic Ocean. We know this because its remains have been found in the mud at the bottom. In the deep, still ocean of the Arctic they sank to the bottom, and there, because the water was so still, there was too little oxygen to allow them to rot. They lay there, with all the carbon that they had acquired from the atmosphere by photosynthesis locked within them.

Although at first it seems incredible, this mechanism alone—known as the *Azolla Event*—may have been enough to account for the rapid cooling that brought the Eocene to an end. There is enough carbon locked in the pickled tissues of *Azolla* at the bottom of the Arctic Ocean to have reduced the CO_2 content of the whole atmosphere—enough to put a stop to the greenhouse world that had prevailed through the millions of years of the Eocene.

But the Azolla Event would have produced such cooling only in the relatively short term. In truth, ever since the Eocene ended, and right up until the time of the modern greenhouse world brought about by human industry, the world has continued to grow cooler. There have been warmer spells and colder spells as the Earth shifted its orbit and moved closer or farther from the sun. But on the whole, the trend has been downward. To explain this, scientists invoke a quite different mechanism—just as fantastical but also, apparently, what probably happened.

This second idea has to do with the chemistry of the air and rocks. As rain falls, it picks up CO_2 from the atmosphere,

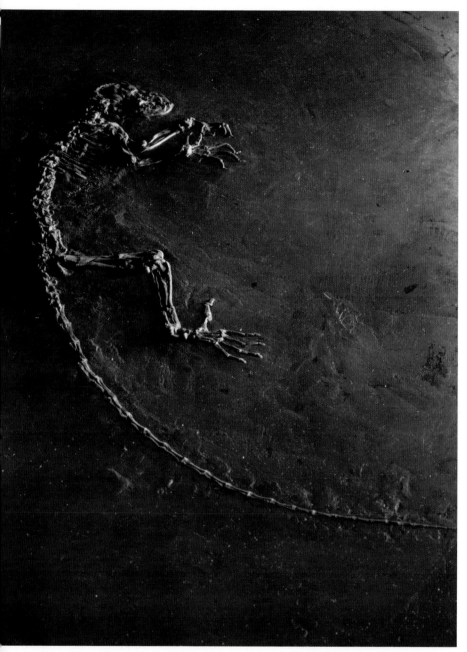

Ida, a 47-million-year-old primitive primate fossil from the Messel Pit in Germany. She is the most complete primate fossil ever found.

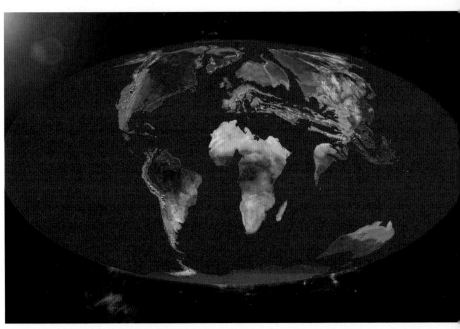

The position of the continents during the Eocene epoch, roughly 50 million years ago.

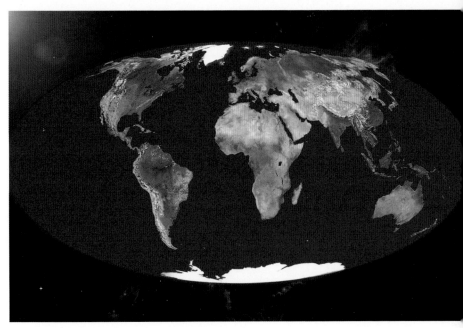

The position of the continents today.

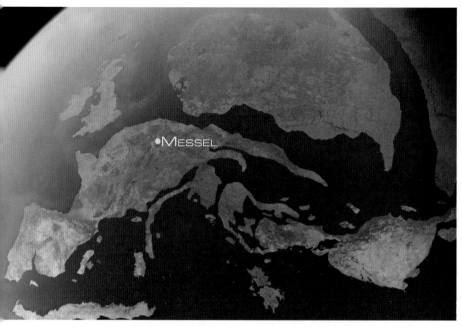

Eocene Europe, roughly 50 million years ago, showing where
Messel, Germany, would have been.

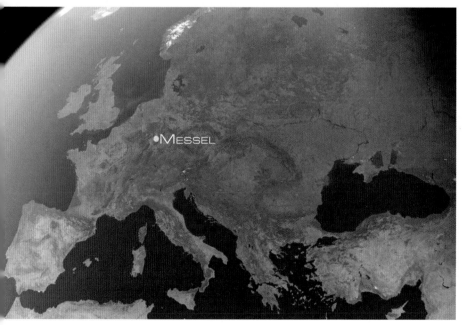

Europe today.

The Messel Pit—a UNESCO World Heritage site, twenty-two miles (35 kilometers) southeast of Frankfurt, Germany. In the Eocene, a rain forest surrounded what was a lake, and the area would have teemed with life.

Fossils recovered from the oil shale in present-day Messel date to 47 million years ago.

Messel's fossils offer dramatic detail and preservation. Here is *Macro-cranion tupaiodon*, an early spineless hedgehog from the Messel Pit.

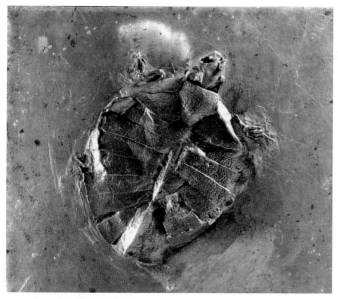

Allaeochelys crassesculptata, an early Messel turtle.

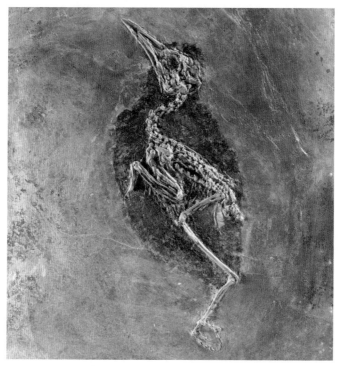

An undescribed bird from the Messel Pit.

Amphiperca multiformis, an early fish from the Messel Pit.

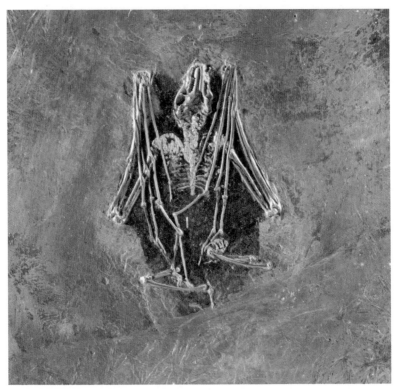

Palaeochiropteryx tupaiodon, an early bat from the Messel Pit.

and so forms a weak solution of carbonic acid. This in turn reacts with rocks such as granite, which are basically made from silicon but also contain some calcium and magnesium. The carbonic acid rain combines with the calcium and magnesium to form bicarbonates of those metals, which are carried away by rivers and into the sea. The silicon that is left in the granite breaks up and becomes sand. The bicarbonates are integrated with rocks in the seabed and eventually are recycled as plate tectonics runs its course. Thus, carbon in the form of carbon dioxide is slowly but constantly circulated between rock, atmosphere, and ocean, and overall remains roughly in equilibrium—an equilibrium that, as things have turned out, keeps our Earth at a wonderfully consistent temperature.

But toward the end of the Eocene, this cozy cycle was interrupted. India is the reason. India had been drifting north from Antarctica for many millions of years, and about 40 million years ago, it finally made contact with the south coast of Eurasia. Then it kept on moving, like a train that has overrun its terminus. It ground into the base of Asia and ruffled the land in front of it as it went. The ruffles became the Himalayas and the Tibetan Plateau. This effect, as Maureen Raymo of the Massachusetts Institute of Technology described it in the early 1990s, was like "a giant boulder thrust into the atmosphere." The rain that fell (and continues to fall) on this giant boulder was prodigious—carried by winds that had crossed the entire length of the southern Pacific and were now thrust upward as they struck the rising mountains. We see the result today in the monsoons. But much of the rain simply flowed, and still flows, back into the sea, and by the mechanisms described above, it carries vast quantities of bicarbonates with it, incorporating carbon from the carbon dioxide of the atmosphere. It carries enough, indeed, to cool the whole Earth, and to go on cooling it, so much so that by the start of the Pleistocene, it was cold enough to freeze over. In the past 2 million years, the world has endured a succession of at least twenty Ice Ages, and we're sure to have another, but not quite yet, given the global warming that we ourselves have caused (and through means that are potentially even more dangerous than these natural events).

So, add the death of the Arctic ferns to the literal rise of Tibet and we have all the mechanisms we need to explain why the tropical days of the Eocene came to an end. But before this period of cooling began and for millions of years, the place we now call Messel was warm and lush, as inviting to a diverse range of species then as it is now for the world's top paleontologists.

CHAPTER FOUR

THE MESSEL PIT

When Ida was alive, during the middle of the Eocene, Messel was much warmer than it is today. How do we explain this? The main reason for this was the prevailing greenhouse effect: much higher CO_2 in the atmosphere. The other reason is that in Eocene times, Messel was a lot farther south than it is today.

To expand on the great Ukrainian-American biologist Theodosius Dobzhansky's comment that "nothing makes sense in biology except in the light of evolution," we could add that a number of evolutionary puzzles themselves, like this one, make no sense except in the light of continental drift: the extraordinary fact that land moves.

Over many millions of years, much of the world's land has moved through many thousands of miles, east and west, north and south, sometimes crossing from one hemisphere to the other, sometimes spinning through 90 degrees or more, sometimes joining up with other masses of land, and sometimes splitting apart. We can see it happening even now. The

Himalayas are rising as the landmass of India continues to shove its way into Asia, and East Africa is splitting free from the rest of the continent along the Rift Valley.

When Alfred Wegener noticed that many of the rocks on the eastern side of South America lined up precisely with similar rocks on the west coast of Africa, he said it was as if a newspaper had been torn, leaving the newsprint still aligned to show how it once was. At first many scientists found the idea very intriguing. Then, as the years passed, more and more agreed that it was implausible. There simply wasn't a mechanism that could explain such weirdness.

But in the decades after World War II, the mechanism became clear. In the windows of ancient cathedrals, the glass is thicker at the bottom than at the top because, over decades and centuries, it has flowed downward. So too continental movement is slow, but over millions of years—and geological time is measured in such intervals—the movement is impressive. As the Canadian geophysicist John Tuzo Wilson suggested in the 1960s, the surface of the whole Earth is divided into about twenty discrete tectonic plates (from the Greek *tecton*, meaning "builder"), which are shuffled by the restless mantle beneath.

In Eocene times, North America was joined to Eurasia by two different routes. To the west of America, what is now Alaska was joined by a wide land bridge to what is now Siberia. To the east, Canada was joined directly to what is now Greenland, which in turn abutted firmly on what is now Scandinavia. The Arctic Ocean was an isolated sea, completely surrounded by land. Animals of all kinds, including land animals, could move freely across the entire Northern Hemisphere. Because the world as a whole was so warm at that time, there was no ice to block their passage, even though the principal thoroughfares between North America and Eurasia were so far north.

Because the position of the continents affects the direction of the ocean currents, the layout of the land is sometimes such that the currents cannot easily flow from the tropics to the poles. When that happens, the poles grow colder, the ice builds up, the albedo increases, and the whole world is cooled. But in the Eocene, water flowed readily from the equator down toward the poles. This warmed the poles and melted some or all of the ice, so the albedo went down and the world warmed up.

The area that is now Messel was about 10 degrees farther south than it is now—roughly on a level with modern-day Sicily. Even today, in this temperate world, Sicily is a great deal warmer than central Germany. In Eocene times, in a distinctly greenhouse world, Messel was very hot indeed. It is not quite accurate simply to call it tropical—*paratropical* is the better term—but it was certainly hot and wet.

The Eocene world's geography was quite different from our world's, and knowing why helps us begin to make sense of both Messel itself in those times and the plants and animals that provided Ida with her habitat and were her neighbors. It also helps us understand how the region came to be the treasure trove of fossils that it is today.

The particular bit of the Eocene that concerns us is now an oil shale pit. But how come? And what is an oil shale pit? What, for that matter, is oil shale? And why is such a peculiar thing to be found at all—and what is it doing in the middle of Germany? Why do fossils last so well in oil shale? How did it ever occur to anyone to look for fossils in the first place? Of what kinds of creatures are the fossils? And why are there so many?

To answer these questions, let's begin more or less at the beginning.

How the Messel Pit Was Created

The Messel Pit is known as a *maar lake*. A maar comes into existence when hot magma—volcanic lava—deep beneath the ground swirls too near the surface and makes contact with the subterranean water table. The water instantly turns to steam, and because of the pressure of earth on top, a mighty explosion occurs—one of nuclear proportions. This creates a huge crater—generally described as shallow but up to a thousand feet (300 meters) deep—with a bank of debris all around. Beneath the crater, reaching down into the ground as far as the magma, is a "chimney"—the exit channel for the hot rocks and gas as they force their way upward from the magma. This chimney can be up to one and a quarter miles (2 kilometers) tall, and after the explosion, the chimney remains packed with pummeled and reconsolidated rock.

The pit at the top of the chimney generally fills up pretty quickly with water—and so the maar becomes a lake. The whole structure in cross section roughly resembles an old-fashioned champagne glass: the magma at the bottom, deep in the Earth, is the base of the glass; the chimney forms the stem; and the lake is the wine in the bowl. The biggest maars are in Alaska, up to five miles (8 kilometers) across. There are others in Texas, New Mexico, South America, and Africa, and among the best known is Lake Nyos in the Cameroon

Highland of Africa. But the first maars were described in the Eiffel region of Germany, which is not far from Messel. The word *maar* is regional German dialect from the Latin *mare*, meaning "sea." The maar at Messel, at its peak, was probably around two miles (3.2 kilometers) across and up to a thousand feet (300 meters) deep.

Clearly, maars happen only in areas with a great deal of restlessness beneath the surface, and Germany is not known for its volcanic activity. But the explosion that produced the maar that has become the Messel Pit occurred more than 47 million years ago, in the early Middle Eocene, when the area that is now Messel was a lot farther south—roughly, as mentioned above, on the latitude of present-day Sicily. Sicily today has Mount Etna and, across the waters in southern Italy, Vesuvius (with the remains of Pompeii right next door, entombed in volcanic ash). In Ida's epoch, the ground beneath Messel was lively.

The general climate in that place and at that time was, as we have seen, warm to the point of tropical. However, it was not quite like the modern tropics. Within the tropics proper, there is no true summer or winter, and the days are always around twelve hours long. In Eocene Messel, just as in modern-day Sicily, there were clear seasons, the days varied in length, and while many trees were evergreen, in the tropics there were many deciduous types as well—the types familiar in higher latitudes, which shed their leaves in winter to reduce water loss and minimize general damage. Thus the climate is best described as paratropical, and although most of the plants were of a kind that is perfectly recognizable today, they tended to occur in combinations that in modern times would seem somewhat strange, like palm trees alongside oaks. As for the animals, some of them would look very familiar to modern eyes and some decidedly would not.

The Ecology of Messel Maar

Lake Messel appeared out of nowhere. It was not the natural center of a catchment area, and for the most part it was filled only by groundwater and by rain. There is evidence that it was fed by creeks from time to time, but probably none of these were permanent. So the water in this very deep lake was also very still. The top sixty-five feet (20 meters) or so would have contained a reasonable amount of oxygen and could have been rich in life, but below that, being neither shaken nor stirred, the water would have been highly anoxic.

With all that forest around—all that teeming life—the lake was surely well supplied with organic nutrients. The upper waters were a paradise for planktonic "algae" of many kinds. Prominent among them was a true alga called *Tetraedron;* indeed, there could have been five hundred million individual *Tetraedron* in every liter of water. As in modern lakes, and particularly in nutrient-enhanced farm ponds, such algae periodically multiplied beyond all bounds to form "blooms"— which probably came and went with the seasons. Whenever the algae bloomed, they soon became overcrowded, robbed the surface water of its oxygen and nutrients, and then died in their billions. Down in the depths, there was virtually no oxygen at all, so instead of sinking to the bottom and decaying in a flurry of heat and dissociated molecules, as they would in a compost heap, the dead microorganisms simply lay, decaying very slowly, compressed by the weight of the water above. Over perhaps one hundred thousand years, and possibly nearer a million, the compressed corpses reached a thickness of up to 650 feet (200 meters) and were transformed eventually into various grades of oil and tar. The details differ from place to place, but in general, with pressure and time, the oily, tarry stuff transformed into various kinds of soft, hydrocarbon-rich black or brownish rock known as shale.

Why Messel Has So Many Excellent Fossils

The forest around the lake at Messel teemed with animals, as rain forests always do. We would expect some of them to fall into the lake now and again. Lakes are enticing and accidents happen.

But there is more to it than that. Lake Messel did not simply entice. Any beast that came too near was liable to be anesthetized and then would drown. Lake Nyos in northwest Cameroon shows at least in principle how this can happen. Magma beneath Lake Nyos leaks carbon dioxide into the water. In August 1986, possibly triggered by a landslide, a huge cloud of carbon dioxide gas rose suddenly from the lake and killed seventeen hundred people in nearby villages, and about thirty-five hundred livestock. Lake Nyos has been made safe by some clever civil engineering, although it still seems precarious. But there weren't any civil engineers in Eocene Germany.

It seems that the Messel maar also puffed out CO_2—at some times more than at others, presumably, but always quite a lot. There were amphibians and reptiles of many kinds in and around Lake Messel. Despite the hazards, there were fish—though generally only the kind that can tolerate very low levels of oxygen. Insects bred in the lake and formed great swarms above it, just as they do over any modern pond. Bats, already well established by the Eocene, flew in to feed off the insects. So did birds. Terrestrial animals of many kinds came to drink. The fossils found at Messel include all such creatures—with fliers in superabundance. As they flew over the lake to hunt and snatch a drink, they simply fell out of the air.

Once dead in the water, these creatures were in conditions that were perfect for fossilization—not just of their bigger, harder bones but also of their finest details. Their bodies lay relaxed because they were drugged and they drowned. Typically, 47 million years after they met their fate, the animals are lying on their sides, as if in some dreamless sleep. They sank into the ooze, which, in effect, pickled them. The flesh is all gone but, often, the shape of it remains in the shale that has formed from the ooze, as an outline, which German paleontologists call the *Hautschatten*, or "skin shadow." We can tell, for instance, whether the animals had long ears or short ears, and often the gut contents have remained in situ, where the intestines would have been.

More than that: we can often see the fine detail of the hair or feathers. This is made possible by an extraordinary concatenation of circumstances. There were some bacteria in the depths—it is hard to find a place on Earth where there are absolutely no bacteria of one kind or another—and in quick time, they covered the entire surface of any corpse that came their way, including every hair and every barb of every feather. After all, to a bacterium, a hair is huge, like an oak tree to a mouse. As the bacteria fed upon the protein of the hairs and feathers, they respired, and as they respired, they produced yet more carbon dioxide in water that was already rich in CO_2. There was also a lot of iron dissolved in that ancient lake. The CO_2 reacted with the iron to form iron carbonate, otherwise known as siderite. The iron carbonate then precipitated out, forming a thin layer over each and all of the bacteria that were coating the hair and feathers. The surface layer of siderite then killed the bacteria. So each hair and feather was beautifully encased in its own individual covering of black iron-based salt.

Even the finest details are preserved. We can see this most spectacularly in the insects. Many of the beetles retain their jewel-like colors. Most creatures derive all or most of their color from organic pigments, which do not generally survive for long after death, even in the best conditions. But insects commonly create the illusion of color by diffraction: the surface is sculpted into an array of miniature prisms that break up the light that falls on them. This is how the elytra (wing cases) of beetles achieve their shimmer. The preservation at Messel is so fine that even the minute surface structure that makes this possible is still intact. In the Eocene insects from Messel, these surface details are preserved, and so too is the iridescence.

Over time, the shale that formed from the ooze at the bottom of the onetime lake was filled with the corpses of animals. The lake was a death trap with a bed of embalming fluid. But how did anyone know that the fossils were there?

By modern times, after all, the shale was covered over. There remained just a depression in the ground, with treasures no one could have imagined buried beneath.

How Did the Fossils Come to Light?

It became apparent in the eighteenth century that deposits of oil shale surrounded the village of Messel, which by then, of course, was situated in its present northern latitude. This was useful fuel in a new industrial age, even without much further refinement, so people started digging for it. In December 1875, the first fossils came to light: 150 fragments of bones, including bits of jaw, from the crocodile now known as *Diplocynodon darwini.*

By then, industrialization was in full swing. The late nineteenth century was the golden age of industrial chemistry, and Germany was a world leader. By now, the finer qualities of oil shale were becoming fully apparent. It wasn't just for burning in a rough-and-ready way. It was concentrated hydrocarbon. Properly refined, it yielded tar, paraffin, petrol, crude oil, and a whole host of other materials destined for various reaches of the new industrial scene. Mining at Messel began in earnest in 1884 as the Messel Mining Company (Gewerkschaft Messel) came formally into being. Messel pit, the erstwhile Eocene lake, became the Grube Messel—the Messel Open Pit Mine.

The miners were not unaware of the fossils. The year 1884 also saw the first geological-paleontological survey of Messel. But for fossil seekers, there was a huge problem. Oil shale, formed underwater, still contains about 40 percent water. As it dries, it crumbles, and the fossils within crumble with it.

If we find a fossil in chalk or some similarly stable stone, we can put it in a drawer for a decade or so and come back when we are good and ready. We cannot do that with fossils found in oil shale. Leave them and they fall to pieces. In the early days, there was no good way of getting the fossils out of the shale and making them secure.

In the late 1960s, the local museum, the Hessisches Landesmuseum Darmstadt, became involved (Hessisches, referring to the region of Hessen, which includes Frankfurt and Darmstadt). The Landesmuseum is one of the oldest public museums in Germany, dating from the late eighteenth century, and it is very highly respected. Mining stopped at Grube Messel in 1971, and the way was clear for serious scholarly excavation. But alas, the Hessen government decided that the gaping pit at Messel would be a very handy place to put garbage. The locals protested, and so did Jens Franzen from the Senckenberg Research Institute at Frankfurt (attached to the very fine Senckenberg Museum). He led the efforts by the worldwide scientific community, but the plans to create a landfill, although delayed by the protests, remained in place for nearly twenty years. Roads were built to accommodate the trucks that authorities were confident would soon bear trash to the pit.

In 1987, following a decision in the courts, the plans for the landfill were at last called off. In 1988, the state court in Kassel declared that the approval given for the landfill was illegal, and in 1991, the state of Hessen bought the Messel Pit for 32.6 million deutschmarks and declared it a natural and cultural heritage site. Only then were the legal battles finally discontinued. The following year, the state government of Hessen and the Senckenberg Research Institute agreed that any fossils that were found henceforth from previously unknown species should remain with the institute or with the Hessisches Landesmuseum at Darmstadt, and in December 1995, UNESCO added the Messel Pit fossil site to its World Heritage List. Inclusion on the list, said the World Heritage Committee, "confirms the exceptional and universal value of a cultural or natural site which requires protection for the benefit of all humanity." Being included on the list is the ultimate recognition. An observation platform was opened in 1997; an information center in 2003; and a permanent center for visitors in 2008 (with guided tours on weekdays and holidays from April through October).

Yet the fossil to which this book is dedicated, a gem even by Messel standards, did not remain in Darmstadt or in Frankfurt. It made its way to Oslo. It was found in the early 1980s by a private collector and was officially legalized when the last mining company gave up its propriety to all fossils coming from their mine. Then the fossil found its way into the trade, to be bought at a Hamburg fossil fair by the University of Oslo's Jørn Hurum. Frankfurt or Darmstadt could have put in a bid, but neither was able to raise the necessary money. In much the same way, German science allowed its other world-ranking fossil to slip through its fingers in the late 1860s — the truly fabulous fossil of the world's oldest bird, *Archaeopteryx* — which was found at Solnhofen and ended up in London. (The Germans did, however, hang on to the second *Archaeopteryx,* found in 1877, which is often said to be even better than the first one and is now in Berlin.)

Amateurs were already pulling out fossils from the Messel Pit in the late 1960s, and although amateurs often get bad press, paleontology owes a great deal to them. In various scientific fields, amateurs have often proved to be technically very adroit and have pioneered new techniques. Professional paleontologists who want to explore some new site have to get official approval from the state authorities and find grants. In the case of Messel, an old treaty between the mining company and the Hessisches Landesmuseum Darmstadt granted all rights on Messel fossils to the Darmstadt Museum. As a result, professionals from other institutes did not get seriously involved at Messel until 1975, after several years of wrangling. But by then there was the imminent threat of the pit's conversion to landfill, and fossil hunters felt they had to work as quickly as possible. Haste is not good in the field of paleontology any more than it is in archaeology. Each site is unique, so there can be no second chances. Almost certainly much has been lost at Messel, especially when the shale was being mined just for its chemistry. But undoubtedly too, there is still much to be found.

A very great deal has been found already, and serious paleontology at Messel became far easier after the 1960s, when a method was finally found to extract the fossils from the crumbly oil shale and embed them for all time in hard but yielding transparent epoxy resin. When we put the fossil knowledge together with what we know of similar regions today, we can see in some detail the creatures that lived there and guess at what other creatures might also have been present.

Life in Eocene Messel

Despite its nasty tricks, Lake Messel clearly harbored a rich assortment of wildlife, and so did the surrounding woods. Ida lived in a very lively neighborhood.

The fossils from Messel include representatives from ninety-six families of flowering plants and a hundred families of invertebrates, plus more than 130 species of vertebrates. These numbers are impressive, but the truth is, they represent only a meager percentage of what must have been present in Eocene times. Now there are more than five hundred families of flowering plants, 70 percent of them occurring in tropical forests, and the Eocene flora can hardly have been less diverse. A single hectare (2.47 acres) of modern tropical forest is likely to contain up to three hundred species of trees and at least forty thousand species of insects. In truth, though, the Messel fossils offer an almost incomparable view of life as it was lived an extraordinarily long time ago and we should be extremely happy with the astonishing array that we do have.

Since maars have steep sides when they form, blasted into existence as they are, trees can grow right up to the water's edge. So, among the fossils at Messel, we find the remains of forest trees, including some giants: various members of the Juglandaceae, the walnut family; more from the laurel family, the Lauraceae; some from the Fabaceae, formerly known as the Leguminosae; and a lot of palms. Walnuts these days are known mainly from warm, temperate climes rather than from tropical forests, but tropical kinds exist too that have small oval leaves with "drip tips" that jettison surplus water like gargoyles off a roof; these resemble the fossil leaves of the laurels found at Messel. And laurel leaves often turn up among the fossilized gut contents of Messel's horses, albeit somewhat scrunched. Leguminous trees too (Fabaceae) feature strongly in modern tropical forests. The palms from Messel are known primarily from their pollen, flowers, and fruits. The leaves of palm trees commonly die on the tree and rot in place. Climbing up and hanging from the many trees would have been plenty of vines and lianas. Members of the Vitaceae—grapevines—are well known from Messel, and they evidently scored highly with the fruit eaters.

Clearly, though, the banks wore away in places and there were also flat and probably marshy areas around the lake, as well as shallows. So we find huge (six-inch, or 15-centimeter) scales from the pinecones of the extinct conifer *Doliostrobus,* which probably grew on the fringes of the lake like present-day swamp cypresses. And there are many remains of Araceae—the family that includes the arums such as the cuckoo pint and the tropical food crop *Colocasia,* known commonly as taro. There are sedges (Cyperaceae) and members of the Restionaceae, which look like rushes but are of a different family; and tupelo trees of the genus *Nyssa,* which now grow in Southeast Asia and the southern United States and are much favored as ornamental garden trees. From the water itself we find the remains of big floating leaves like water lilies, though whether they were close relatives of modern water lilies is unclear.

Although Messel's climate was effectively tropical, it was still seasonal, and more temperate trees grew nearby. Among the remains at Messel is the fossilized pollen of beech and pine. Beyond a doubt, the habitat was varied, offering many different niches; even if the lake itself did have a nasty tendency to drug and drown its visitors, it was perfectly hospitable in its upper reaches and in the occasional shallows, at least for much of the time, and we would expect to find a commensurately huge variety of animals. The fossil record offers only a tiny glimpse, but it is enough to suggest that this was indeed the case.

Puzzles with Animals

Though the forest of Eocene Messel was not quite like any modern forest, botanists would not find much in it that they did not recognize or could not at least ascribe to a modern family. But a zoologist unversed in paleontology and unsure of what to expect would surely be at sea. There were plenty of animals there that no longer exist. It isn't just the occasional species that has disappeared. Entire zoological families or even orders from the Eocene have gone, leaving no modern descendants—nothing alive to compare them with, to help make sense of them. And there were many more animals at Messel from groups that do still exist today but that were very different from their modern counterparts. Then there are those that have been found from species and families that are still with us but no longer survive in Europe. Often—and it is very strange indeed—we find that the nearest living relatives of Messel's Eocene animals now survive only in South America. North America is not known to have been fully connected to South America until about 3 million years ago, in Pliocene times. The odd species might have rafted across the Atlantic from Europe or Africa to the coast of what is now Brazil or Argentina, but mass exodus of entire menageries is altogether implausible. If all those animals walked, what was their route? They might have walked into North America via one of the northern points of entry and then, somehow, rafted or island-hopped down to South America—although if they did that, they sailed against what was then the prevailing current. They might have walked across the Isthmus of Panama once it appeared, but some of them are known to have lived in South America before the isthmus was created, at least in its present form. In any case, if the present-day denizens of South America got there via North America, why, in many cases, have they left no trace in North America?

In passing, there is a lesson in all of this for modern zoos. Today's zoos tend to be very fond of zoo geography. Instead of keeping monkeys with monkeys and big cats with big cats, they tend to prefer to keep them all in themed communities: the savanna of Africa, the rain forest of Indonesia, and so on. This is all very well. Animals do live in communities, and the communities do form integrated ecosystems, where each has adapted to the presence of all the others. Remove any one of them and all the rest must adapt or go extinct. Add one from elsewhere—an exotic—and again, all the rest must adapt or die. It matters where animals and plants live, and it can be extremely damaging to transport any creature of any kind from its own established niche and dump it in some other country.

But it is a mistake for zoo visitors to assume, for example, that just because elephants and lions now live in Africa and tropical Asia, that is where elephants and lions are bound to live. Fossil elephants of many kinds have been found all over Europe and in both North and South America. Today's Asian elephants apparently arose in Africa. Lions never made it to South America, but they too in their time have all but covered the globe. Rhinoceroses, now confined to Africa and Asia, originated in North America.

It sometimes seems as if most animals in their time have lived in most places. Even for animals that are not given to rapid migrations, distance is no barrier. In a million years, a lineage of creatures could circumnavigate the globe a hundred times over, even if its individual members moved at tortoise pace. Climate is not necessarily a problem—not, for example, when extreme latitudes are temperate, as they were in Eocene times. Oceans, glaciers, deserts, and mountains can get in the way, but these, as we have seen, tend to come and go; and, of course, birds, bats, and many insects can fly over them, and land animals can sometimes get across by a variety of creative methods, for which, again, there was plenty of time.

Overall, it begins to seem that the country where any particular animal lives now is simply where it happens to have ended up. Certainly all the creatures in any one place will form a community with the others that are there, but if the coin had flipped differently, then all but the most extreme specialists could just as easily have formed communities with some other suite of neighbors. This does not mean that we can give ourselves carte blanche to relocate the wild animals and plants of our present world as if it doesn't matter, as in practice we have often done. But it does mean that when we find fossil creatures in places that to us now seem bizarre, we should not be too surprised. Paleontologists need a very broad search image, and even then it can still be hard to explain how any one creature or group got from where it was in the deep past to where it is now. And the ancient route from Eurasia to South America remains a puzzle.

One last problem. We identify creatures from the past by comparing them with the ones from the present. If a creature's bones are like a modern monkey's, we call it a monkey. If they are like a parrot's, we call it a parrot. If it is obviously a mammal but not like any modern kind, we may give it a new family or order. It is all very logical. But there is many a slipup. Rarely do we have the complete skeleton, and odd bits can be very deceptive. Just as deceptive, often, is the phenomenon of *convergence*. Creatures that have different ancestors but adopt the same way of life tend to evolve along similar lines and end up looking alike even though they are unrelated. Thus, it should not be possible to mistake a primate, but the earliest kinds in general form do look remarkably like squirrels. And there is also *divergence*. Creatures that are closely related but adopt very different styles of life will evolve along quite different lines, and although evolution is supposed to move slowly (Darwin emphasized "gradualism"), when the pressure is on, change can be rapid. Closely related animals can soon look very different. So, to see who really is related to whom, taxonomists tend to look not at the most obvious feature — like whether the teeth are adapted for eating leaves or grass, since this is the kind of thing that can change rapidly depending on way of life — but at minute details of the kind that do not need to change according to the way of life. They look, for example, at the precise placements of the exits through the braincase of the cranial nerves. This is why paleontology can often seem so esoteric and so concerned with the little things.

Though the Tertiary period is often called the Age of Mammals because it is the first great age in which mammals had a chance to show their potential, many other groups that were already well established when the Tertiary began have also flourished and diversified wonderfully — and many animals besides mammals have given rise to entire new families and orders even since the Tertiary began. Mammals didn't fizzle out during the Tertiary, and Messel shows this very well. The mammals are wonderful—and of course include the heroine of this book—but still, among the fossils they are very much in the minority.

Messel's Nonmammal Species

Insects and Fish

At Messel, there are insects you would expect to find in Europe, and some you wouldn't, and there are some that look like those of today, but when we look at them closely, they prove not to be the same. The fossils include a large cicada and a very large bush cricket known as a katydid. There are thirty-six specimens of "walking leaf"—related to stick insects but disguised in an almost perfect imitation as leaves. The ones at Messel are more or less identical to present types; but today they live mainly in southern Asia. There are mantidflies, predatory relatives of the lacewings, which catch small insects by folding their front legs in the manner of a praying mantis (which is unrelated). Nowadays, mantidflies live in South America. There are also termites. There are three kinds of bees of a type that live in colonies like modern honeybees, but two of those ancient lineages are now extinct. Evidently they could not survive the switch of climate at the end of the Eocene. There is an extraordinary ant, *Formicium*, the largest ant known: the queens were almost an inch (about 2.5 centimeters) long, with a wingspan of up to half a foot (16 centimeters). They too were killed off by the cooling that followed the Eocene. All of the above must have been blown into the lake from the surrounding forest. There are aquatic insects too, which presumably lived in the shallows at the edge of the lake. These include scavenger beetles; caddisfly larvae; and phantom midges, which presumably fed on and among the floating algae.

Plenty of fish have come out of Messel. Some of them are modern bony fish of the kind known as *teleosts,* which literally means "ultimate bones." This is the group that includes salmon, cod, and goldfish. But most of the fish from Messel are bowfins and gars, which come from an altogether more primitive group of bony fish that are traditionally called Holostei. The holosteans were at their peak in the Mesozoic, the great dinosaur age, but one species of bowfin and seven species of gar still exist in the lakes and rivers of North America.

Gars and the bowfin look seriously primitive, particularly the gars. Gars wear heavy body armor like chain mail, formed from interlocking bony scales, and their scales are further reinforced with ganoine, which is basically the same as the enamel on the teeth of mammals (and it grows in the same way). Two species of gar are known from Messel. The commoner one, *Atractosteus strausi,* looks like the living gars, with a heavily armored and sculpted head and a protruding snout with a row of sharp teeth like a crocodile. Clearly, like modern gars, it was a serious hunter of other fish.

But the commonest fish by far among the many hundreds found at Messel is a bowfin, *Cyclurus kehleri,* which is very similar to the surviving North American type. The Messel bowfin was armored too and had a massive skull. Like its surviving relative, it was clearly a fierce predator, mostly of other fish—which means there must have been plenty of other fish for it to feed on. But it surely flourished as well as it did at Messel because, like other holosteans, it was able to breathe air. When it gets too stuffy in the depths, bowfins and gars can gulp air directly into the swim bladder, the wall of which has a heavy blood supply for the exchange of oxygen, like a lung.

Fewer teleosts are known from Messel. This might be because there really weren't as many there, but it also might be because they do not fossilize so easily. In general, teleosts have dispensed with armor. Their scales are lightweight (and so, usually, are their internal bones), and some kinds almost do without scales altogether. Like Greek heroes, they seem content to go naked into battle.

There are also three species of perch. The order to which perch belong—the Perciformes—is extremely successful and diverse in the modern world, and between them, the three kinds at Messel show some of the order's versatility. To judge from its teeth and general shape, one species was a highly maneuverable hunter and ambusher—the inshore equivalent of a leopard. Another species seemed built for a leisurely cruise through open water, probably browsing on plankton. The third has crushing plates at the back of its throat, which points to its having been an herbivore, but comparison with similar fish today suggests it actually crushed small animals.

There is even an eel—only one has been found—of the genus *Anguilla,* very similar to the European eel of today. Eels can migrate over land up to a point if it is damp enough; and they also migrate from salt water to freshwater. Its presence at Messel is one of several clues that the lake may not have been quite so cut off from the outside world as is generally supposed—or at least not all the time. Most probably this eel got there by migrating upstream.

Amphibians, Reptiles, and Turtles

We might expect to find plenty of frogs, toads, salamanders, and newts at Messel. The kinds that live mainly on land would surely have relished the damp undergrowth and soil in the surrounding forest, and all amphibians (except for a few modern extreme specialists) need to lay their eggs in water. But few have been found. Perhaps the water didn't suit them. *Eopelobates* was there, however—a long, strong-legged jumper. It seems to be related to some modern frogs from South America (again we see the South American connection), but it may also be ancestral to the present-day spadefoot toad of Europe known as *Pelobates*. Rather than being a jumper, the spadefoot is a digger. Also at Messel is another frog, from a family that is now extinct, and one salamander.

Reptiles, in contrast to the amphibians, lay their eggs on land, even though some of them spend much of their time in water and feed in water. All the modern reptile groups were present in Eocene Messel: snakes and lizards, crocodilians, and turtles (including some that the British usually call terrapins).

Lizards and snakes are closely related. Lizards evolved very early, way back in the Permian period, 250 million years ago, well before the first dinosaurs. The snakes evolved from lizards in the late Cretaceous, only about 65 million years ago. By paleontological standards, snakes are modern—much more modern than the mammals. The jaws of snakes and lizards are specific to them: the different bones can come apart so that they can open their mouths extraordinarily wide in order to, as some of them do, swallow other creatures that are fatter than themselves. They also have scales (the two groups together are called the Squamata, meaning "scales").

There aren't many fossil lizards from Messel (most lizards have no particular need to go near a lake), but those that are there belong to five different families. They include legless and armored lizards, with their bony-plated heads and shoulders, from the family Anguidae; skinks, of the Scincidae; wall lizards, in the same family as the modern European common and sand lizards, Lacertidae; one from an extinct family mournfully known as the Necrosauridae, meaning "dead lizards"; and an iguanid, from the Iguanidae. Except for a few oddball outliers in Madagascar and the Fijian islands, all the Iguanidae today live in the Americas.

The twenty or so snakes at Messel include two small constrictors related to modern boas, which nowadays come only from Central and South America; and a pipe snake that may be related to the modern coral snake and that probably burrowed into loose earth on the forest floor.

The turtles are an ancient group that still thrives throughout the world—on land, in lakes, and in the sea—even though overall there are only around three hundred living species. Of all the fossil vertebrates found at Messel, they are among the most common, and since turtles as a group will not breed at temperatures below 77 to 86 degrees F (25 to 30 degrees C), they confirm (if confirmation were needed) that Messel was warm. Some of the fossil turtles were in pairs, apparently copulating; and since turtles like to copulate in shallow water, this again confirms that the Messel maar wasn't all steep-sided.

The Messel turtles include representatives from five genera

of the Cryptodira—the kind that can draw their head right into their shell; and one genus from the Pleurodira—the "side-necked" turtles, which simply flip their head sideways to get it out of the way. The cryptodirans include *Emys*, close relative of the modern European pond turtles; and *Allaeochelys*, which seems transitional between a pond turtle and a soft-shelled turtle and is closely related to the modern Pig-nosed Turtle that lives in New Guinea (which, given the apparent exoticism of so many Messel creatures, is not a particularly surprising address). The biggest of the Messel turtles is *Trionyx messelianus*, with a shell up to two feet (60 centimeters) in length. *Trionyx* now live in the tropics and subtropics. Like frogs, these turtles can exchange oxygen through the soft skin, where the shell does not obtrude. They feed mainly on fish. The only side-necked turtle from Messel, *Neochelys franzeni*, is small and fairly rare. It is the oldest Tertiary side-neck found in Europe (there are older ones, but they date from the Cretaceous) and the first side-neck to turn up in Germany. Perhaps side-necks disappeared from Europe between the Cretaceous and the Eocene, or perhaps the ones that were there have simply eluded the paleontologists.

Crocodiles like it hot, and their presence also affirms the warmth of Messel. They too are an ancient group—from 230 million years ago and again predating the dinosaurs—but while all the existing kinds of crocodiles walk on four legs and are at least semiaquatic, the very first kinds walked and ran on two legs, and many in the history of the world have been more or less exclusively terrestrial. By Eocene times, the bipedal ones were all extinct, but there were still plenty of landlubbers left. Some of the species known from Messel were aquatic and some were terrestrial.

Eight species of crocodiles from six genera are known from Messel. This is a surprisingly high number from one small area; in modern times, there are rarely more than two kinds of crocodilian in any one place. All the Messel kinds were specialist feeders (whereas some of the modern types tend to be more generalist), and creatures that feed on different things can often live side by side without too much trouble. It's probable, though, that the Messel crocodilians did not all live in or around the lake at the same time.

Only two of the eight species from Messel are thought to have lived in the lake at least some of the time. The others would have been visitors, or temporary residents. Commonest among them was the one that was found first—*Diplocynodon darwini*. The name means "double dog tooth" and refers to the protruding tusklike fangs in both jaws, which were in the same position as a mammal's canine teeth and clearly visible even when *Diplocynodon* had its mouth closed. *Asiatosuchus* was a large beast—up to sixteen feet (5 meters)—with a massive skull suggesting strong jaws. It was a powerful predator, like a modern Nile crocodile, able to attack large animals. But it seems to have been more like modern alligators (which today live in the southern United States and in China).

Less common were two species of "grinder" crocodile of the genus *Allagnathosuchus*: one only four or five feet (1.2 to 1.5 meters) long, and the other even smaller and slimmer. These were more closely related to alligators. They had short snouts and broad, blunt teeth at the back that look as if they were good for crunching mollusks. The modern African dwarf crocodile has similar teeth but eats fish and frogs—and some fruit, which seems a most uncrocodilian thing to do. There was also an armor-plated type less than three feet (1 meter) long called

Baryphracta deponiae, which also had blunt teeth at the back of the jaw and so far is known only from Messel. Then there are a couple of saw-toothed crocodilians that presumably were just passing through. One of them, *Pristichampsus,* was clearly a carnivore (why else have teeth serrated like steak knives?), but its claws were converted into miniature hoofs, so it was clearly a land runner. There was also a gharial, like the modern gharials, with a long, thin snout ideal for catching fish. These are now hanging on in India and perhaps—just barely—in a few other places in Asia.

Birds

A modern birdwatcher in Eocene Messel would have been utterly entranced and almost utterly confused. All of the birds seem to belong to orders that still exist—those of the cranes, the kingfishers, the nightjars, and so on. But many of the Eocene birds require their own families, with no modern counterpart; and many of the Eocene types look nothing like the modern species, even when they do belong to the same order; and a few look nothing like any kind of bird that lives today. And as with any group, convergence and divergence make it even harder to decide who is related to whom.

Some modern birds are conspicuously absent from the Messel lineup. On land these days we expect a high proportion of passerines—perching birds, like finches and thrushes. After all, the single order of the Passeriformes contains more than 60 percent of all living species of bird. But the passerines, like the grasses, did not truly come into their stride until well after the Eocene—to some extent, the rise of the passerines depended on the rise of grasses—so there are no passerines at Messel. Other small birds filled the niches that the passerines now tend to dominate. Particularly prominent before the passerines came fully into their own were the Coraciiformes—kingfishers and rollers—and there are certainly some at Messel.

So far, Messel has yielded no ducks. Yet ducks were widespread in Eocene Europe. Perhaps there are no ducks from Messel because ducks are particularly light and might have floated long enough to decay on the surface, never sinking into the preservative ooze below. But in truth their absence is a mystery.

Hundreds of skeletons from more than fifty species of birds have been identified from Messel. Many of them did not live on the lake, and many of them were clearly flightless, so they could not have flown over it. Instead, they lived as modern pheasants and jungle fowl do, in the woods all around. But they still managed to drown themselves.

The commonest bird fossils of all are from *Messelornis cristata,* known colloquially as the "Messel rail." It was about the size and general shape of a moorhen, while the *cristata* refers to a horny or fleshy comb on the top of its head. It may not seem at all odd to find such birds in a lake, since rails in general, including moorhens, are aquatic birds. But the Messel rail had long legs and short toes, suited neither for swimming nor for striding over mud as a moorhen can nor for trotting along the lily pads like a jacana. Its wings were short, so if it flew at all, it was at best a flutterer. Probably it was not truly a rail but closer to the Sun Bittern, which now lives only in South America. So, now we have a double mystery: Why should a land bird drown in such numbers in a lake? And how, despite its inability to fly, could it be related to a bird that now lives on a quite different continent?

Though we have only one of its bones from Messel Pit—a femur found a very long time ago, in mining days—the biggest of all the Messel birds was *Gastornis,* which stood more than six feet (2 meters) high and yet was stocky, weighing in at 220 pounds (100 kilograms), with a head as big as a modern pony's and a huge eaglelike beak. Here the American connection is very strong, for *Gastornis* seems to be more or less the same as the American *Diatryma,* which was first described from New Mexico in 1874. *Gastornis* has the general mien of a megapredator, with its huge bulk and terrifying beak. Birds descended from dinosaurs, and in *Diatryma/Gastornis,* we see the dinosaur heritage shining through; it is like a miniature (though still impressively large) avian *Tyrannosaurus rex.* Perhaps all the world would now be dominated by these reincarnated dinosaurs had it not been for the rise of the cats, wolves, and bears. If we look more closely, however, we see that *Gastornis*'s beak seems better equipped for scything vegetation, and its legs are not built for running, nor its feet for killing. It may well have been a formidable opponent, as many an herbivore can be (including buffalos and elephants), but it was not a specialist predator.

There are many other birds too with the general form of bustards or cranes—flightless or almost so. We again see the South American connection with *Palaeptos weigelti,* which resembles a modern bustard but is possibly related to the rhea; and yet again in *Salmila robustus,* a massively boned bird about the size of a ptarmigan that resembles the South American seriemas and trumpeters. *Juncitarsus merkeli* is the size of a crane, with very long legs, and it is apparently related to the flamingoes and grebes (which, strange though it seems, are apparently closely related to each other). Oddest of all, *Juncitarsus merkeli* is the only true waterfowl of any kind found at Messel. As already noted, there are no ducks. There is, however, *Rhynchaeites messelensis*. It is called the "snipe-rail" because it has features in common both with rails and with painted snipes. But the structure of the beak tells us it is related to the ibises.

There are hundreds of finds from small arboreal birds too—everything from impressions of single feathers to entire skeletons still articulated and with a complete set of feathers. In some it is possible even to see how the colors were patterned, if not the colors themselves. Some birds were very small; some indeed were probably hummingbirds. Most of the small ones ate insects and fruit, though there is perhaps the odd nectar feeder. Dedicated nectar feeders, such as modern hummingbirds, need specialist beaks, and it seems that the ancestors of hummingbirds honed their skills on insects. *Paragornis* had qualities of both swifts and hummingbirds: it fed on insects but was small and short-winged and perhaps a hoverer, like a modern hummingbird, and it gleaned insects from the undersides of leaves. *Scaniacypselus,* on the other hand, was a very definite swift, which could easily be mistaken for a modern swift. *Gracilitarsus mirabilis* may well have been a specialist nectar feeder, but it may not have been a hummingbird at all. (The full story can be very confusing, even to specialists!)

Among the many birds you wouldn't expect to find in Eocene Messel are *Paraprefica kelleri,* which are related to the modern potoos. They spend their days sitting stock still on branches, pretending to be a broken branch themselves; and these days, yet again, potoos are exclusive to South America. The "Messel hoopoe," *Messelirrisor,* is a distant relative of modern hoopoes. *Primozygodactylus* is like a modern woodpecker, with two toes pointing forward and two backward—excellent for gripping—and it seems to be a true woodpecker. But the contents of its stomach show that it ate grapes. There are some modern relatives of woodpeckers that also specialize in fruit. They are the toucans. There was a parrot too at Messel, which had a straight, pointed beak and looked nothing like a modern parrot.

There were also avian predators around the lake. There are two skulls of a hawk about the size of a large sparrowhawk. And there were big predators, including two species of owl, both about the size of modern barn owls and quite big enough to menace small primates.

And then there were the mammals.

The Mammals of Messel

The mammals of the Eocene fall into three general categories.

In the first are those that were doing well in the Mesozoic and survived the K-T catastrophe but disappeared around 35 million years ago, in the Early Oligocene. Only one group fits this category—the multituberculates, so-called because their teeth had so many cusps. The multituberculates were small and rodentlike, and, presumably, in the end the rodents outcompeted them.

In the second category is a broad suite of orders that evolved and diversified rapidly at the start of the Tertiary, when the dinosaurs had left the field. By the Eocene, many of these orders were at their peak. These early but long-gone types include several that are spectacular. The uintatheres were huge-hoofed herbivores the size and shape of a modern rhinoceros, with knobby faces and huge fanglike tusks on their upper jaw. They must have been as formidable as rhinos, but they barely survived the Eocene, and they have no close living relatives. The mesonychids included some paradoxical creatures that had hoofed feet and yet were obviously carnivores. They included the biggest land carnivore that ever lived—*Andrewsarchus*—which was about twice as long as the biggest modern bears and probably weighed in at around three quarters of a ton. Skulls of *Andrewsarchus* have been found, and its head was clearly nightmarish: somewhat wolf-like but long-jawed like a crocodile's and, of course, huge. It did not survive the Eocene. The mammals found at Messel have all been much smaller than a uintathere or the biggest mesonychids. These giants are mentioned here only to emphasize that by the time we get to the Eocene, well into the Tertiary, mammals were already enormously diverse and some were very big.

Thirdly, though many of them died out without descendants—and many entire families have disappeared—we find Eocene versions of the kinds of mammals that are still with us, and some of them must be the direct ancestors of today's mammals. All the modern orders are known from the Eocene, though many if not most of the modern families came on board later. For instance, the modern order of the Carnivora was well in place by the Eocene, but within that order, the particular families of the bears and the hyenas did not appear until much later. The primates too were flourishing in the Eocene, with Ida among them, but our own family, the Hominidae, dates from only about 5 million years ago.

No multituberculates have been found at Messel, at least not yet. But there are plenty of mammals from the other two broad categories. Many have gone extinct, with no descendants and, in many cases, no obvious close living relatives. There are several that clearly are Eocene versions of creatures that are still with us, although many of the modern descendants no longer live in Europe or anywhere near. Again, at least one of them is now known only from South America. It's as if South America were a global refuge camp for animals that had in other places fallen by the wayside.

The Mammals from Orders That Are Long Extinct

The mammals of Eocene Messel were Ida's neighbors and the creatures that her species had to adapt to—some as rivals and some as potential allies.

Hunting within the lake itself was the otterlike *Buxolestes*. It belonged to a very primitive group of placental mammals known generally as the Proteutheria. Bits and pieces of *Buxolestes* or creatures very like it are known from North America, but Messel yielded the first complete skeleton. There were two species at Messel. The bigger one (found in 1977) was about two and a half feet (80 centimeters) long, of which about 40 percent was tail; and the places where the muscles inserted show that the tail was powerful and used for rowing. Where its gut would have been are fish scales and an entire fish jaw—and bits of other vertebrates too. Like a modern otter, *Buxolestes* could undoubtedly hunt on land as well.

Yet it was definitely not an otter. Otters, quite independently and much later, adopted the same shape and modus operandi. Nature is endlessly inventive, but it endlessly reinvents nonetheless.

A second species of *Buxolestes*, 20 percent smaller, turned up in the Messel Pit in 1988. Its teeth point to an otterlike diet. But the fossilized contents of its gut suggest that it ate plants. Nature never ceases to surprise.

Another proteutherian predator from Messel was the extraordinary *Leptictidium*. Among modern creatures, there is nothing quite like it. It was a giant leaping rodent that hopped along on its hind legs and was roughly the size and shape of a modern spring hare. But it had a pointed head, and the places where the muscles inserted suggest that it had a short trunk, like a modern elephant shrew. It probably did not hop as efficiently as a modern kangaroo, but its long legs suggest that it jumped well enough, and its stomach contents include the remains of small lizards and a big insect, possibly a grasshopper. It was an agile carnivore of the forest floor. It has been suggested, incidentally, that kangaroos and their relatives first evolved their ability to jump not in the open plain, where they live now, but in the forest. When the undergrowth is thick, hopping can be a good way to get through.

Then there was *Kopidodon*, resembling a giant squirrel more than three feet (1 meter) long, though 60 percent of its length was a big bushy tail. Unlike a squirrel, but like a primate, it had an opposable thumb and big toe to give it a good grip—another example of nature reinventing a winning formula. *Kopidodon* may have been an herbivore but it probably had omnivorous leanings.

Messel's final representatives of this broad group of pro-

teutherians are four specimens of *Heterohyus,* known colloquially as "longfingers." They were small—a mere foot (30
centimeters) long, of which half was bushy tail—and their
large feet suggest that they could hop about in the trees like a
modern bush baby. But they seem to have fed like a modern
aye-aye, with chisel-like teeth for breaking into trees, and a
long finger (in fact, two long fingers on each hand) for pulp-
ing the insect larvae that lurked within and scooping out the
flesh. Aye-ayes use the middle finger; *Heterohyus* used the
second and third; and *Dactylopsila,* a marsupial from New
Guinea of similar size and shape, does the same thing using
its elongated fourth finger. But this is not the end of the convergence. The most adept wood chiselers and scoopers-out
of insect larvae are the woodpeckers, and all of these three
mammals live in countries where there are no woodpeckers,
as is the case with Messel.

There was a creodont at Messel too. Creodonts as a group
look like the modern Carnivora, the order that includes the
cats, dogs, weasels, bears, and so on. But the creodonts were
quite separate. Here is yet another wondrous example of con-
vergent evolution. The easy way to tell Creodonta and Carnivora apart is by their carnassial teeth—cheek teeth with
sharp edges for shearing meat. Carnivora have only one carnassial tooth on each side, on the upper and lower jaw, and
the other cheek teeth are more modest. Creodonts commonly
had two carnassials on each side of the upper jaw, and another
on the lower jaw. It's been suggested that this made it difficult
for creodonts to vary their diet, so if meat was not available,
they went hungry. But it's clear that some creodonts ate some
fruit, and among modern Carnivora, the cats are committed
meat eaters. Nature is never simple.

Messel's credont was *Lesmesodon*. It was small and agile, and the fossil remains are good enough to tell us that it had a thick, bushy tail. It was very much like a squirrel, except that it probably lived mainly on the ground, where it would have occasionally run into Ida.

Mammals of a Kind Still with Us

All the rest of the fossil mammals from Messel probably belong to orders that do still exist.

Four species of marsupial are known from Messel (though there are only six specimens in all). Everyone knows that marsupials these days live in Australia; they also live in New Guinea and in South and North America (which has several opossums). Europe may come as a surprise. But the fossil evidence shows that in their time, marsupials have lived in all the continents. Three of the Messel marsupials seem to have been terrestrial, but one—provisionally called *Paradectes*—lived in the trees, as many marsupials do. *Paradectes* was small—a mere ten inches (26 centimeters), of which two-thirds was the prehensile tail—and clearly an agile climber.

The most puzzling find of all at Messel is a possible anteater, *Eurotamandua,* its name derived from a modern arboreal anteater called the tamandua. Anteaters are placed in the order Xenarthra, together with armadillos and sloths. As a group they seem quite distinct from other mammals because their vertebrae have a unique structure, which includes an extra point of articulation (the name Xenarthra means "hidden joint"). Xenarthrans clearly came into their own in South America, even though the earliest-known record is from Messel, and quite a few of them made it into North America (where in the west, armadillos are standard roadkill, just as hedgehogs used to be in Europe).

Eurotamandua certainly looks like an anteater. It has no teeth, and its long tubular skull presumably housed a long, tentacle-like tongue. It had limbs that could surely break open a termite nest. But there are no remains of ants or termites in its gut—though there are traces of stuff that looks as if it could have been part of a termite nest. If it really is an anteater, it would be the only complete one ever found outside South America, and at 47 million years, it would be by far the oldest one ever found anywhere. The oldest otherwise known anteater dates from the Early Miocene, around 24 million years ago. But on anatomical grounds, *Eurotamandua* raises doubts. Some paleontologists feel that it may in truth be closer to the modern pangolins (sometimes called "scaly anteaters" but from a quite different order). Time and more finds might clear up the mystery. In any case, before Ida came along, *Eurotamandua* attracted more attention from professional paleontologists than any other fossil from Messel, extraordinary though many of them are.

Messel does have some true carnivores—members of the order Carnivora—though they seem to belong on the margins of the modern group. (The general term *carnivore* describes any creature that eats flesh, from crocodiles to eagles to lions. Carnivora is a distinct group, a true taxon, which nowadays includes the cats, dogs, bears, and so on.) Best known of the Messel types is *Paroodectes*, which had the size and shape of a modern domestic cat and, like a cat, was clearly an agile hunter of small creatures. But it was not a cat. Here we see convergence again.

There were plenty of rodents at Messel. Rodents' front teeth are self-sharpening (the top ones rub against the bottom ones at just the right angle), and if they didn't gnaw, the teeth would never stop growing. As such, they are the supreme gnawers, and by the Eocene they were becoming the principal occupants of small ecological niches, seeking out seeds on the forest floor and in the trees but also feeding more eclectically—modern porcupines are great consumers of bones—and soon occupying a wide variety of habitats. Nowadays, with about two thousand known species, the single order Rodentia includes almost half of the living species of mammal (the bats, with almost a thousand, account for another quarter). But their numbers are not merely a matter of biological "success." In part, they are so numerous because there is more room in the world for small animals than for big ones.

The biggest rodent found at Messel is *Ailuravus macrurus*, like a truly massive squirrel, three feet (1 meter) in length. But it wasn't agile, and apparently it was a leaf eater. *Masillamys beegeri* was substantial too, at sixteen inches (40 centimeters) long, and shaped like a vole. It looks as if it could climb but probably lived mainly on the ground. The smallest of the known Messel rodents was *Hartenbergeromys parvus*, which was the same general shape as *Masillamys* but only about a foot (30 centimeters) long, of which about 40 percent was tail; it was roughly the size of a rat. There was also a dormouse, *Eogliravus*. This was a true dormouse of the family Gliridae, which is still going strong. *Eogliravus* resembled the modern fat dormouse, also known as the edible dormouse because the Romans used to eat it.

Quite a few insectivores have been found at Messel. In

truth, as with *carnivore,* the term *insectivore* when applied to mammals can cause some confusion. Again, *insectivore* with a small *i* simply means "insect-eater"; but the term is also applied to a discrete order of mammals that traditionally was called Insectivora and today includes the hedgehogs, shrews, and moles. But in fact, many primitive mammals are or were insectivorous whether they were closely related to one another or not. Because they were or are primitive, they tend to have very few distinctive features of the kind that enable taxonomists to decide who exactly is related to whom. It is doubly hard to decide whether any ancient mammal, possibly known only from a few bones and teeth, is actually a relative of modern shrews or simply in a generalized way primitive. To some extent the problems are being sorted out; for instance, the formal group Insectivora, which used to describe the modern shrews and so forth, is now called Lipotyphla. A primitive anatomy and a penchant for insects are not enough to qualify for membership in this group.

But the Messel fossils do seem to include bona fide members of the modern Lipotyphla, or at least close relatives thereof. These include six specimens of *Pholidocercus,* which was about sixteen inches (40 centimeters) long and had bony scales on its tail, a horny crown on its head, and long, thick bristles along its back. Presumably it rummaged for what it could find on the forest floor, like a modern porcupine.

The most common of all the land-bound mammals found at Messel (more common than anything except bats, that is) was the hedgehog relative *Macrocranion tupaiodon,* about a foot (30 centimeters) long, with a pointy face and mobile snout surrounded by long tactile hairs; short, woolly fur; long ears; and small eyes. Clearly it relied on touch, hearing, and smell rather than vision, which are typically nocturnal traits. *Macrocranion tupaiodon* had long, strong hind legs and apparently could spring—presumably to escape from predators. The contents of its gut show it was an omnivore, with a taste for fish, which it presumably found in the shallows, since it was clearly no swimmer. Ecologically, indeed, it seems more like a raccoon than a hedgehog. There is also a half-size and much rarer relative of *Macrocranion tupaiodon* called *Macrocranion tenerum,* which seems to have been a dedicated insect eater on the forest floor, and perhaps was even more of a leaper when alarmed, like a modern jerboa.

Before Ida, the most famous of all the Messel fossils were its perissodactyls: two extraordinarily intact tapirs and—above all—a supreme array of ancient miniature horses. (The other remaining group of perissodactyls are the rhinos, but there are none at Messel.)

The first of the two tapirs was found in 1973, after almost a century of systematic mining. The earliest known tapirs date from about 54 million years ago, from North America, but the Messel skeletons are the oldest that are still complete. The Messel ones are of the genus *Hyrachus,* which may or may not be the same as the Eocene kinds from North America. Today tapirs live only in Central and South America and in Southeast Asia. The Messel tapirs are not tiny, but they were very small by modern standards—about five feet (1.5 meters) long—as the new wave of Eocene mammals all tended to be. Some zoologists see their diminutiveness as an adaptation to the tropical or subtropical heat—since small animals find it easier than large animals to keep cool. However, Philip Gingerich has suggested that the high CO_2 content of the Eocene atmosphere had a stunting effect. Plants thrive on high CO_2, but animals on the whole may not.

Messel's Eocene horses are extremely tiny. More than sixty complete skeletons have been found at Messel, together with a great many bits and pieces collected since the first horse was discovered in 1911. The whole of North America, by contrast, has produced only one complete Eocene horse. There are stallions, mares, and juveniles, and eight of the mares are pregnant.

Eocene horses had four toes on each forefoot and three on each hind foot. Messel's horses are from three genera. *Propalaeotherium* was the first one to be found; it stood about twenty-one inches to two feet (55 to 60 centimeters) at the shoulder. Then *Hallensia* came to light; and finally, *Eurohippus*. *Eurohippus messelensis*, the kind now found most commonly at Messel, was only thirty-one inches (80 centimeters) long and a mere twelve to fourteen inches (30 to 35 centimeters) high at the shoulder—not much bigger than a large domestic cat. Apparently its back was arched, like that of a modern muntjac (a deer from Southeast Asia) or a duiker (an African antelope). The teeth of these ancient horses were not tall and ridged like a modern horse's but smaller and more rounded. Evidently, as befits a forest animal, they were mainly browsers, feeding on tree leaves, but in the gut of one of them are grape seeds. The high-crowned smashing and shearing teeth of the modern horse, as well as its great size and one-toed feet, are adaptations to grassland; but grassland did not become extensive until the warmth and wet of the Eocene had come to an end. (The change of climate and the decline of forest are highly pertinent to even our own evolution.) There were also a few, but very primitive, artiodactyls with long tails at Messel—the order of the even-toed ungulates, which now include deer, cattle, sheep, and antelopes. But they are very rare and obviously living solitary lives. Yet the artiodactyls are now the commonest and most diverse ungulates by far.

The mammals that turn up most often at Messel are the bats. The oldest-known bona fide bats are 53 million years old, but there is some evidence for them from deep in the Cretaceous. Most likely the dinosaurs would not have worried them, but they still had to compete with birds, which were well established in the Cretaceous, and indeed with the flying reptiles, the pterosaurs. Presumably they specialized from an early stage as nocturnal aerial hunters of insects. They would have had that niche more or less to themselves.

By the mid-Eocene, bats were successful and diverse, and there are thousands of specimens from Messel. Although most fossil bats from elsewhere are reduced to fragments—because bat bones are fragile—the ones from Messel are mostly complete. In general they are similar to modern bats, although their echolocation was less sophisticated, which we know from the structure of their inner ears, sometimes preserved in fine detail. We can tell from the shape of their wings and from their size relative to the body that some of the bats maneuvered extremely well but flew quite slowly and hunted near the ground, like a modern horseshoe bat; others hawked for insects between the trunks and crowns of trees; and still others, narrow-winged and built for endurance, flew above the canopy. Modern bats carve up the ecosystem in just the same way. Yet most of the bats found at Messel belong to families that no longer exist. The only exception is Trachypteron.

Why are there so many bats among the fossils at Messel? One answer, surely, is that there were a lot of bats about. Another is that bats often like to fly over water, partly to snatch a sip and partly because it's a rich hunting ground for insects. It seems that over time, many were overcome by gusts of CO_2 and simply fell out of the sky and down into the bed of oily preserving fluid deep beneath the lake.

Primates

The primates must also have been common around Messel. The paratropical forest was just their kind of territory, and by the mid-Eocene, they too were diverse. But there are only eight specimens of primates from Messel, which between them come from three species. Two of those species are from the lemur-like genus *Europolemur*: *E. koenigswaldi* and *E. kelleri*. The remaining one is Ida.

Why are so few primates preserved at Messel when there are, for example, so many horses and bats? One reason is their way of life. Small primates live mostly up in the trees. Rain forest trees trap plenty of drinking water one way or another, and their inhabitants possibly had little reason to come down to the ground at all. Some of the primate fossils that have been found suggest that they did not necessarily die without assistance. One piece of lemur—a lower jaw—was found in a coprolite of *Buxolestes;* and in one of the lemur bones, a crocodile tooth is embedded. Perhaps the croc found it dead and carried it in.

All of this makes the discovery of Ida the more miraculous. Here she is, the only one of her kind yet known; conceivably the only one of her kind that ever fell into the lake; perfectly preserved and one of the world's most extraordinary fossils; and indeed the best-preserved fossil primate ever found of any age. Yet despite Ida's rarity in the record, there was a lot going on in the primate line, and there had been at least since dinosaurs roamed the Earth.

CHAPTER FIVE

HOW PRIMATES BECAME

Ida and her relatives from the Eocene belong to the grand order of mammals known as the primates—as do lemurs, bush babies, monkeys, and apes, and as, indeed, do we. Primates are an extremely varied lot: at least 260 species are known, and biologists are still counting. Many of our fellow primates are small, live in dense, tropical forests, and emerge only at night, making them very hard to study, and several of them are only now becoming known. So the real number is probably nearer three hundred. There could be almost as many kinds of primates as there are kinds of parrots (also midsize animals that live mostly in tropical forests).

Not all primates are located directly on the human evolutionary chain, but they all play a role in how we understand the combination of characteristics that have made us the most developed of the primates. Radical transitions in primate evolution occurred throughout the Eocene, from 56 million to 34 million years ago. Many scientists argue that the primates that were in the direct line of humans must have lived during the Eocene in sub-Saharan Africa, but exactly what kind of primates those would have been is not known because there are huge gaps in the fossil record. This is where studying Ida in her entirety and with a view forward opens up a new chapter in primate evolution. Just as Ida complicates primate history, she gives us hints of where a transition occurred in the great story line of primates, because she allows us to see a combination of complex primate traits all in one skeleton. In general, after all, the basic characteristics found in Ida are present in primates that lived in the ensuing 47 million years.

Primates come in all shapes and sizes. The smallest of the living primates, the pygmy mouse lemur, is the size of a mouse—a mere two and a half inches (6.2 centimeters) in length—and weighs in at a little more than an ounce (30 grams). You would get about fifteen of them to the pound. But alpha male gorillas, the biggest living primates, may reach more than 330 pounds (150 kilograms) in the wild (and a great deal more in captivity if they are badly fed and get fat), meaning that a big male gorilla is about twice as big as a man and generally about three times as strong. The biggest living primates, even when they are not obese, weigh at least five thousand times more than the smallest. If mouse lemurs were puffed up to the size of a man, then gorillas on the same scale would weigh about four hundred tons—about four times the probable weight of the *Mayflower*.

Different kinds of primates look very different from one another, move very differently, and live very different lives. The most familiar types (including ourselves) have big domed heads and relatively small noses (and many apes and monkeys have protruding jaws), but the more primitive kinds have long snouts like a raccoon (although the word *primitive* needs teasing out). Most live in tropical forests much like Ida's home, but a few monkeys and one kind of ape—the kind known as *human*—have strayed into temperate zones, in our case virtually to the poles. Most kinds live in trees, some in the highest canopies, but some climb trees only to rest or for safety, and quite a few live more or less full-time on the ground.

Up in the trees, some run along the branches, some hang underneath, some leap spectacularly from tree to tree, some swing with long arms, and some help the acrobatics along with long, muscular tails that serve as a fifth limb (though apes and some monkeys have no tails at all, or only tiny ones). Among those living on the ground, most walk on four legs all or most of the time, but a few walk on two legs at least some of the time. In truth, only one living species of primate—or of any animal—looks truly convincing when walking on two legs with the body held vertically: *Homo sapiens*. Our own ability to walk upright—not to mention to walk downstairs while carrying a tea tray and thinking about something else, or to dance the samba—is truly miraculous. The only other modern animals that walk habitually on two legs are birds, but they cheat, maintaining their balance with a head and neck that stick out in front and a tail that sticks out behind, like a tightrope walker's pole. True, penguins contrive to walk with body upright, but on the whole they make an ungainly mess of it (and when the snow is smooth, they prefer to flop on their bellies and punt themselves along with their wings). Human beings, walking perfectly erect on relatively small feet, are intrinsically unstable, like a pencil standing on its tip. It's only our wondrous coordination, an internal telephone exchange of reflexes in tendons and brain and everything in between, that keeps us standing proud.

Primates vary socially too, enormously. Some are monogamous. Many are polygynous (one male with several "wives"), and a few are polyandrous (one female with more than one "husband"). Many live in groups centered on one or several males, and some live in groups centered on the females. Some stay in groups more or less all the time, while some prefer to be on their own at least when they are feeding. Most, including all of the Old World monkeys and the apes, are diurnal, meaning they are active only or mainly by day, but the bush babies and many others of the more primitive types, plus a few New World monkeys, are nocturnal. Social living is easier by day, when the individuals can communicate by sight, than by night; the nocturnal ones tend to be more solitary. All the different aspects of an animal's life are interconnected.

But if the primates are so diverse, why do we place them all in the same zoological order? What do they all have in common? What makes a primate a primate? How can we be sure—and there can be no doubt about it—that Ida and lemurs and monkeys and, indeed, human beings, all belong to the primate club?

What Makes a Primate a Primate?

Taxonomists prefer to define each discrete group according to one or a few simple and invariable characters, that is, features that are shared by all the members of the group but are not found in other groups. Thus, among mammals, it is easy enough to define bats, in the order Chiroptera. All bats have leathery wings formed from the bones of their fingers, with powered flight, and no other mammal has anything like this. Xenarthrans (alias edentates)—the group that includes anteaters, sloths, and armadillos—are also easy to identify. The xenarthrans vary enormously in shape and size, but all have characteristically and peculiarly shaped neck vertebrae, which suggests they shared a common ancestor. Perissodactyls, including horses, tapirs, rhinos, and the strange extinct chalicotheres, have hooves and bear their weight mainly on their middle toe. Modern horses have *only* the middle toe, and although there were plenty of three-toed horses in the past, they too bore their weight on the middle one.

But with primates, there are no such simple, infallible defining features. There are features that all primates have in common, and such features, as a philosopher would say, are necessary, because if an animal doesn't have them, it can't be a primate; but they are not sufficient defining features, because some animals that are not primates also have them.

Some primates do possess features that no other kind of mammal has. But not all primates have these very special features, so we say that these features, when they appear, are sufficient to define a primate but not necessary. There is no single feature that is both necessary *and* sufficient by which to define a primate, but if we put a whole "suite" of features together, they do allow us to see when an animal is a primate and when it is not, and to make a good guess at whether it is or isn't, even when we have nothing to go on but a fossil bone or two. Again, having a nearly complete skeleton makes this task much easier.

If we see an animal with fingernails, we can say that it must be a primate. But this is one of those "sufficient but not necessary" features. Marmosets—small tropical American monkeys—have claws on all their fingers and toes with a nail on the big toe only.

All primates have a collarbone. But so do many other mammals, so a collarbone is a "necessary but not sufficient" feature. We can say that an animal lacking a collarbone cannot be a primate, but we can't say that one that has a collarbone is definitely a primate.

The eyes are a big feature of primates. Most mammals rely very heavily on their sense of scent. Dogs are outstanding among the many beasts that run their entire lives according to what they can sniff: they can tell who has passed their way; which females are ready for sex and which emphatically are not (a good thing to know when intended mates have very big teeth); where there is food; and so on. Human beings are far more olfactory than we generally give ourselves credit for, and aftershave laced with pheromones has become big business, but on the whole we rely most heavily on our vision. So do most primates. In fact, as primates have evolved over the past few tens of millions of years, there has been a general shortening of the face and a rising of the forehead, emphasizing the eyes (and brain) at the expense of the protruding raccoonlike snout. The very first primates had teddy bear faces with protruding muzzles—and modern lemurs on the whole still do—while we and most apes and monkeys (with the exception of baboons) have round heads and snub noses. Even so, many of the first primates (and the modern lemurs and bush babies and some others) do have very big eyes—and still have muzzles too.

Primate eyes have some special, distinctive features. They see in color, which most mammals do not, and the eyes point forward so that the fields of vision overlap, allowing them to see in 3-D. For creatures that live in the trees, as most primates have since they first appeared on Earth, this seems very desirable. They need to judge distance, which can be done without stereoscopic vision by moving the head, but stereoscopic vision is best, and if you are leaping about 150 feet (50 meters) above the forest floor, you can't afford too many mistakes. All bona fide primates also have a postorbital bar—a strut of bone behind the eye. This means that the eye socket is completely encircled by bone, giving their skulls a goggle-eyed look. Anthropoids take this one step further. Instead of a mere bar of bone, they have a complete bony eye socket. But again, some other mammals also have forward-facing eyes, including domestic cats, and some, such as cattle, do have complete bony spaces for their eyes. So here are yet more features that are necessary to identify a primate but not sufficient.

More distinctively primate, but still not unique, is the opposable thumb and/or the opposable big toe. *Opposable* means that the thumb (and/or big toe) can be swiveled around so that it presses toward the palm of the hand (or the sole of the foot); and with a little refinement it confers *precision grip,* which means that the thumb can grip against any one of the other four fingers. The grip is useful for climbing and for gathering fruit, and when it comes to making tools, as people and chimpanzees do, the precision grip is invaluable. As we will see, the interplay of hand and brain is probably what prompted the brains of humans to so spectacularly outstrip those of other apes, and of course of all other mammals. But marmosets do not have opposable thumbs. Perversely too, it seems, some of the most athletic arboreal primates of all, such as the spider monkeys of South America, do not use their thumbs to grip the branches, which is what you might think thumbs would be for. Instead, they use their long fingers as hooks and allow their thumbs, now reduced to stubs, to simply stick out at the sides. This demonstrates that evolution never stops. A lineage of animals develops a wonderful anatomical device for getting around in trees (an opposable thumb) and then finds some even better way of getting around and abandons the very organ that seemed to be the answer to any arborealist's dreams. On the other hand, squirrels in particular make great use of their forefeet and come close to having an opposable thumb. So even the opposable thumb, which is so distinctive in us and in most apes and monkeys, is neither necessary nor sufficient to define a primate beyond question.

Then there are details of soft tissue: two breasts on the chest; a pendulous penis; and testicles held in a scrotum (and in male mandrills, the penis is bright red and the scrotum is lilac—reminding us that primates are emphatically visual creatures and in general the most colorful of mammals). Without such features, brightly colored or not, a mammal cannot be admitted to the ranks of primates. But elephants and sea cows are among the other mammals that also have two breasts on the chest (as opposed to having a rear-mounted udder like a cow, or a row of teats like a sow); and pigs are among the other mammals that have a scrotum. So these features again are necessary but not sufficient.

Primates also have three distinct kinds of teeth—incisors, canine teeth, and molars (or we could divide the molars into molars and premolars and say four types). Many other mammals have very different arrangements, such as elephants, which just have up to four giant molars in the mouth at any one time and two incisors in the upper jaw in the form of tusks (or sometimes no tusks at all); or killer whales, which have rows of simple teeth like pegs. But many other mammals also have three kinds of teeth, again including pigs. So yet again, the complex teeth of the primate are necessary but not sufficient.

There seems to be surprisingly little emphasis in most accounts of primates on the absolute importance of their mobile arms. No other creature uses them as modern apes can do, to cling, to swing, to grasp, with one hand or two. Cats are remarkably agile in a general way, but no cat can whirl its arms through almost every angle of a sphere as even little old ladies routinely manage to do in their exercise classes. If the arms were not so mobile, the hands would not be nearly so useful. A horse can move its feet only forward and backward—and even if its "hands" were as dexterous as the finest seamstress's, it still could not bring them together to thread a needle. But not all primates can whirl their arms quite so convincingly as an ape or a human. So here is yet another feature, distinctive though it is, that is sufficient and yet not necessary.

However, although there is no one feature that can pin down a primate once and for all, the suite of features taken together does. No other animal except a primate has forward-pointing eyes *and* two pectoral breasts *and* an opposable thumb. It's the combination that does it; the combination that puts the tiny mouse lemur in the same camp as the gorilla, and indeed as ourselves.

To truly understand who Ida is and was and how she might have lived—where she fits in the grand company of primates—we should look at the primates that are still out there, which we can study firsthand. To do this, we need to use taxonomy—the classification of creatures into groups, using art, craft, and science.

Primates in Their Place

Modern taxonomists like to classify living creatures according to their *literal* relationships: whether or not they have common ancestors. Thus, it is assumed that all species in the genus *Homo*—including us (*Homo sapiens*) and the Neanderthals (*Homo neanderthalensis*) and what in the old days was called Peking Man and all his kind of the species *Homo erectus*—descended from the same creature, himself a bona fide *Homo* who lived a little over 2 million years ago, probably in Africa. All the living primates are perceived to belong to the same group because they have particular features in common, as described above, but it is also assumed that they all descended from the same ancestor (who probably lived way back in dinosaur times and resembled a tree shrew).

Groups of creatures that all share a common ancestor are called *clades*. A clade can be small, with only a few members, like the genus *Homo;* or very large, like the primate group as a whole. In an ideal taxonomy, little clades nest within bigger clades which nest within even bigger clades. So the mammals as a whole form a clade of which the primates as a whole are a subgroup.

Clades have to be defined carefully. A clade is not a true clade unless it contains *all* the descendants of any particular ancestor plus the ancestor itself; and it must not contain any other creatures that are not part of the lineage. Thus, we should not arbitrarily declare that squirrels belong to the clade of the primates just because in a certain light they roughly resemble some lemurs. Squirrels belong with other squirrels in the squirrel clade—which in turn belongs with mice and porcupines and others in the grand clade of the rodents. No one has ever suggested that squirrels should be given honorary status as primates. But it is very easy to make mistakes, especially when all you have to go on are a few fossil bones. The *ideal* is to define a nested hierarchy of clades. But the ideal can be difficult to achieve. DNA studies help a great deal, but they cannot usually be applied directly to creatures that are long extinct. For extinct creatures—which may or may not include the ancestors of living types—we have to rely on fossil bones, and usually we have only a few.

In practice, however, biologists who are not specialist taxonomists also divide up living creatures according to what are called *grades*. Thus, in any one lineage of creatures, we find that some species closely resemble the ancestor, and those that do are said to be *primitive*. Some of the primitive types resemble the ancestor because, in fact, they lived just a short time after the ancestor itself and have had little time to become different. But often we find creatures that are alive today that closely resemble the fossil types that we know are from the deep past, and then they too can be called primitive. Among primates, lemurs are said to be primitive because they are presumed to have retained many of the features of early primates. On the other hand, modern grasses and spider monkeys are said to be highly *derived* because they have clearly changed a great deal since their early ancestral days. The terms *primitive* and *derived* are not value judgments.

Primitive does not mean inferior or incompetent. Indeed, the most primitive modern creatures are in some ways the most successful because, quite clearly, they have remained on this Earth more or less unchanged for a very long time. The wonderful hand of human beings that enables some people to play Beethoven violin sonatas is more primitive in some ways than the hoof of a horse because in general shape at least, it is very similar to the five-fingered hands and feet of the world's first terrestrial vertebrates (the general shape of hand that we also see in modern lizards). Similarly, it is better to use the neutral term *derived* than the term *advanced*, which does sound like a value judgment whichever way you look at it.

Among primates, biologists have commonly recognized three distinct grades, and three distinct clades. We know very little about the most primitive ancestral grade of primates—though we can guess that the first-ever primates lived in dinosaur times and that they resembled modern-day tree shrews. Both of the other two grades of primates are still with us, and going strong. The most primitive of the bona fide copper-bottomed unmistakable primates are called the *prosimians*. The other—and most derived grade—are the simians, otherwise known as the *anthropoids*.

The prosimians and anthropoids are then divided vertically into three clades, each of which is a true ancestral lineage. Two of these clades are of prosimian grade; and the remaining clade contains all the anthropoids. One of the prosimian clades includes the lemurs and all their relatives, and the other one includes the tarsiers. The anthropoids (both a grade and a clade) include all the monkeys and apes.

With these principles of taxonomy to work with, we can

get a feel for all primates just by looking at the fairly short list of grades and clades.

The first of the prosimian clades of primates is the *strepsirhines,* and it includes the lemurs, pottos, lorises, and bush babies. These all have the general hallmarks of primates, but they also have a few very distinctive features, one of which they share with their fellow prosimians, the tarsiers. In both groups, the frontal bone (the bone of the forehead) and the mandible (the lower jaw) are in two halves, divided by suture. This is the primitive condition, which we would expect to find in the first-ever ancestor of all the primates. In anthropoids, including ourselves, the two halves of the primitive frontal bone and of the mandible are firmly fused, so our frontal bone and our lower jaw are each now one big solid bone. Strepsirhines and tarsiers also have an elongated second toe on the hind foot (the one equivalent to the index finger), which forms a grooming claw. All mammals and birds need to keep their fur and feathers in the best possible condition and their skin as free as possible from parasites—and this distinct piece of anatomy suits that very purpose.

But strepsirhines also have features that their fellow prosimians, the tarsiers, don't have. In strepsirhines, the eardrum is surrounded by a ring of bone that is not fused to the skull. In anthropoids, this bone is fused to the skull. The arrangement in tarsiers looks intermediate between the two. In strepsirhines, the upper lip is split, like a rabbit's (whereas a tarsier's top lip is intact, as is ours). Crucially, in strepsirhines, the rhinarium—the area around the tip of the nose—is moist and glandular, as in a cat or dog. The nose of tarsiers (and monkeys and apes) is dry. In lemurs too the bottom incisors are sharp and point forward. These are sometimes used to chisel through the bark of trees to get at the sap or the insect larvae beneath and sometimes as a comb, again for grooming. Some of these kinds of features are obvious, and some hide under the skin, but biologists depend on them to sort out the relationships of ancient fossils, although it is very rare indeed to have anything resembling soft tissue to work with. It is these minutiae that can help us decide who Ida really was, and where she fits on the grand family tree.

Prosimians

Lemurs

The strepsirhines as a whole are conventionally divided into seven families, and five of those families are known as lemurs. The families of lemurs are all very different from one another—probably as different as cats are from hyenas—but they all seem to have a common ancestor. Authorities differ markedly on how many species of lemurs there are; some say as few as forty (although that is surely far too conservative), and some say more than ninety. Much depends on whether the biologist regards different types as true species or merely as races.

But however many species of lemurs there are, they all live in Madagascar today, and it seems they always did. Madagascar is a huge island—the fourth-biggest island on Earth, virtually a minicontinent—and it has been an island for at least 125 million years. It seems more or less certain that the ancestors of lemurs—not yet true lemurs themselves—evolved in Africa and arrived in Madagascar around 40 million years ago. Apparently they crossed the Mozambique Channel, which divided Madagascar from Africa, by rafting on floating vegetation. This might seem unlikely, but it happens. Every now and again substantial chunks of vegetation, complete with roots and soil, break off from riverbanks with all their inhabitants and passing visitors on board, and drift out to sea, and every now and again these precarious vessels with their precious cargoes float all the way to some new continent. The hazards are clearly enormous, but nature has millions of years to play with, and over such time the most unlikely things can happen.

Somewhere between ten and twenty of the lemur species form the family of "typical lemurs"—the Lemuridae. The Lemuridae include such stars as the ruffed lemurs, which live up in trees; and the ring-tailed lemurs, which, uniquely among modern lemurs, live mainly on the ground, marching around on all fours in troops, with their black-and-white-striped tails held aloft like an old-fashioned barber's pole and serving as signals, like a tour guide holding up an umbrella.

The eighteen or so recognized species of the Indriidae include the biggest living lemurs of all, about seventeen pounds (up to 7.5 kilograms), the indri and the sifaka. The eight or so species Sifakas live in troops of up to thirteen animals, and although they are basically tree dwellers, they have a wonderful way of leaping sideways along the ground on their hind legs, holding their arms above their heads for balance. When they are not cavorting in this way, they spend a great deal of time sunbathing. Indris are endearing creatures with long woolly fur that live monogamously in small family groups. Like the howler monkeys of South America, they announce their presence to the world—and their command of their territory—from the tops of trees with an eerie chorus that can be heard for miles. In indris, as in ring-tailed lemurs—and among lemurs in general—the female is the boss. While she browses high in a tree on the tenderest leaves, her mate sits in the branches below. If he ventures up before she is good and ready, she bats him on the nose in lemur fashion, and down he goes again to wait his turn.

The extraordinary aye-aye has a family all to itself (the Daubentoniidae). It is about the size of a domestic cat and in general size and shape is like a red panda. Most peculiar, though, are its hugely elongated middle fingers. It listens in trees with its batlike ears for the larvae of insects that lie beneath the bark. It taps the bark with its long middle finger to elicit the telltale movements. Then it chisels into the branch with its protruding front teeth and inserts the elongated digit like a probe, pulps the larvae within, and scoops them out. The Malagasy people have an ambiguous attitude toward the aye-aye. For some it is sacred—if they trap one by mistake, they anoint it with precious oils before releasing it. Others see the aye-aye as a witch—if it points its middle finger at you, you die—and they kill it on sight. Despite this, and the ever-dwindling forest, it survives, largely due to serious conservation efforts.

The sportive lemurs (of the family Megalapadae), around seven species of them, seem paradoxically named. The first part of their name—*mega*—means "large," but in truth this family includes some of the smallest species. They also seem far from sportive, since they spend much of their time sitting around. This, apparently, is to save energy, for they are mainly leaf eaters, and leaves do not provide a huge amount of energy. Small animals, which waste a lot of energy because they lose body heat so quickly, generally need richer fare—insects and fruit.

The last of the groups that are still in existence are the dwarf and mouse lemurs, perhaps thirteen or so, in the family Cheirogaleidae. They include the smallest of all the primates, which, as befits their small stature, are mostly fruit eaters, but they also eat insects and the gum from trees. Fork-marked lemurs are gum specialists, with protruding teeth to chisel holes in bark, and long tongues to ream out the sticky nutrients beneath.

There are lots of lemurs, but we know that within the past fifteen hundred years, at least a dozen more species have gone extinct. Most of them were bigger than the biggest of today's lemurs, and a few were positively huge. One, called *Palaeopropithecus,* about the size of a German shepherd, hung from branches, climbing slowly like a modern-day sloth. *Archaeolemur* was a ground dweller, roughly the size and shape of a modern baboon. The biggest of all, *Megaladapis,* was bigger than a modern gorilla, which is one of the biggest primates of all time, and almost certainly was a ground dweller too, perhaps climbing now and again, as a modern grizzly bear might do.

So what went wrong fifteen hundred years ago? The answer is that the first people arrived. For 40 million years, the ever-growing, ever-diversifying lineage of lemurs had had Madagascar more or less to themselves. They did share it with some seriously huge tortoises and the heaviest birds that ever lived—the giant elephant birds, somewhat shorter than a modern ostrich but very stocky, and probably weighing around half a ton. They got along with these creatures well enough, but human beings made short work of the biggest lemurs—the ones that were easiest to catch—and the elephant birds and the supergiant tortoises too. Madagascar is huge, but it is an island nonetheless, and island species are particularly vulnerable. The populations are limited and they have nowhere to run, and in addition, island creatures that have never been exposed to human beings tend to be too tame for their own good. Most of the lemurs that are left are vulnerable too, even though they are smaller and more elusive. Without serious conservation, they would surely follow their cousins into oblivion.

Lorises, Pottos, and Bush Babies

Outside of lemurs, there are two other families of strepsirhines, the Lorisidae family and the Galagonidae family. The Lorisidae family consists of seven species of lorises, pottos, and golden pottos—the lorises in tropical Asia and the pottos and golden pottos in Africa. The lorises and the pottos are climbers, gripping the branches very tightly and deliberately, and some—known as slow lorises—move extremely slowly, like actors feigning slow motion. The bush babies, on the other hand, are truly prodigious leapers (of which more later). Yet they are similar enough anatomically to be placed in the same suborder; and they share the features that show them to be strepsirhines. Their frontal bone and mandible are each divided into right and left, and they have the characteristic grooming claw. As is typical of small animals, they are basically insect and fruit feeders, and they are nocturnal.

The Galagonidae family consists of the bush babies, or galagos, of Africa. Bush babies illustrate spectacularly how little we really know of primate variety. Even a few years ago, most standard texts told us that there were only half a dozen or so species. An excellent general account of mammals published in 2001 tells us that there are seventeen. But it seems like Simon Bearder and his students discover new species of bush baby every time they venture into the forests of West Africa. Dr. Bearder suggests that the total number is probably nearer forty. Because bush babies are nocturnal, they identify one other by their calls. It was impossible for scientists to identify them all properly until they had the right equipment to make recordings in the field; and they could not distinguish definitively between the different types until modern science enabled them to study their DNA.

Tarsiers

The last group of living prosimians are the tarsiers. Five species are known, all living in various countries in Southeast Asia: the Philippines, Sulawesi, Borneo, and Sumatra. Tarsiers look like bush babies and they live similar kinds of lives, but in detail they are very different. They share the definitive feature of all prosimians: their frontal bones and lower jaws are still in two halves, separated by a suture.

But perhaps the most characteristically tarsier-like feature of the skeleton is the heel bone, or calcaneus. In human beings and other creatures that walk on flat feet, the calcaneus serves simply to turn the corner between the leg bones, which are upright, and the foot bones, held horizontal. In tarsiers, after the calcaneus has turned the corner, it is extremely elongated—which effectively gives the tarsier another joint in its leg. The hind legs of tarsiers, like those of bush babies, are already long—good for leaping—and the elongated calcaneus makes it even longer. All these things, however, are a matter of degree. Fossil prosimians are known that have longer-than-usual calcaneus bones, but they may not be close relatives of tarsiers. They may have acquired this feature independently.

Tarsiers have anthropoid features too. The nose of the tarsier is dry; the ring of bone around the eardrum is fused with the skull; and the upper lip is not split. So, although the tarsiers are of prosimian grade, they have often been placed in the same clade as the anthropoids. But some primatologists say that tarsiers are not close relatives of anthropoids—that the similarities are superficial. As we will see, however, Ida makes it both easier and harder to decide who is related to whom. It is great to find new fossils, especially when they are as brilliant and as old as Ida, but they don't always solve all the problems.

Clingers and Leapers

Being a clinger and leaper is one last, very important feature that tarsiers share with some strepsirhines. A *clinger and leaper* is a creature that clings to tree trunks with its body upright and then thrusts against the trunk with its powerful hind legs and hurls itself into space. With the body still held vertically, it twists in the air, its long bushy tail providing balance, and grasps the next tree in line.

Among the strepsirhines, sifakas and indris do this; and so, spectacularly, do bush babies. Unless we assume that the common ancestor of tarsiers, lemurs, and bush babies was also a clinger and leaper—which would mean that the first-ever bona fide primate must have been a clinger and leaper—this means that each of these three groups must have evolved this ability independently. Clinging and leaping is a very peculiar and very effective way to get around the forest, and though not all primates do it, it is a characteristically primate thing to do.

And the leaps are prodigious. Tarsiers, which are only a few inches long—typically about half the size of a grey squirrel—can leap nearly fifteen feet (4.5 meters) from a standing start, about forty times their own body length. Bush babies achieve the same. Sifakas are said to be able to leap up to thirty-three feet (10 meters) from a standing start. By contrast, only two human beings in the recorded history of the world have ever jumped more than twenty-nine feet: Bob Beamon in 1968 and Mike Powell in 1991. Both, however, took tremendous run-ups (human long jumping is largely an exercise in sprinting), and Beamon, at the Mexico Olympics, was altitude-assisted. As long jumpers, sifakas are in a different class.

We can tell whether or not a particular primate is a clinger and leaper from its skeleton. In most primates, all four limbs are roughly the same length, so most have the rough shape of dogs when they walk on four legs. Some monkeys and apes of the kind that swing from branches—spider monkeys, gibbons, and orangutans, and chimps to a lesser extent—have spectacularly long arms. But all the clingers and leapers have very long hind legs. In tarsiers, the thigh, lower leg, and foot are all roughly the same length, so the three sections together are like a spring; and the hind limb, including the foot—which includes the elongated calcaneus—is about twice as long as the body and head together. The hind legs of sifakas and bush babies are not quite so long, but they are significantly longer than the arms. So when paleontologists find a new primate fossil—on the rare occasions when they have anything other than a bit of skull or tooth to look at—they can tell a huge amount about how it moved, or didn't move, from the proportions of its bones.

In Ida's case, her measurements show that she used her comparatively long hind limbs for leaping to the next tree and that she could sit upright with her body vertical, meaning that she was a clinger and leaper.

The Anthropoids

The last great group of primates are the anthropoids. They are divided into two main clades: the New World monkeys, known as the platyrhines, from tropical America (from Mexico down deep into Chile and Argentina); and the Old World primates, also known as catarrhines, from Asia and Africa (and still with a toehold in Europe). The catarrhines in turn are divided into two groups: the monkeys, which generally, though not always, have tails; and the true apes, often known as anthropoid apes, which include ourselves and don't have tails.

Apes and Us

Including human beings among the true apes raises problems of nomenclature. In the olden days, taxonomists split the living groups into three families. They placed the eleven known species of gibbons, which they commonly called *lesser apes,* in their own family, the Hylobatidae. They placed the *great apes*—the chimpanzee and bonobo, the two species of gorilla from Africa, and the two species of orangutan from Borneo and Sumatra—in the family Pongidae. And they then put human beings and our immediate relatives (all of them extinct) in their own family, Hominidae. This is the classification shown in most textbooks.

The trouble with that arrangement is that the families of animals are supposed to represent real relationships. As most people know these days, modern human beings and modern chimps are very closely related, with around 99 percent of DNA in common. The gorillas are only slightly more different from us—and the chimps are closer to us than they are to gorillas. The orangutans, on the other hand, are clearly less closely related to African great apes than the African apes are to one another or to us. So it makes no sense to put the orangutans in the same group as the African apes and then draw an arbitrary line between the African-apes-plus-orangutans and ourselves.

Modern taxonomists tend to extend the meaning of the term Hominidae to include not only the humans but all the great apes as well, down to and including the orangutan and several extinct types. But this raises another set of problems. First, even though zoologists are supposed to be strictly rational scientists, many of them retain the notion that human beings really ought to maintain some distance between themselves and all other animals. They feel that human beings ought to have their own family, no matter what. They would like to cling to the old meaning of Hominidae, defined as "us and our immediate and extinct relatives and nothing else."

Those of us who don't mind being linked so closely to the chimps and gorillas face a different kind of problem. Members of the Hominidae are informally called hominids. There is now a huge literature on human evolution, and nearly all of that literature uses the term *hominids* in the traditional sense: to mean members of the genus *Homo,* plus a few closely related groups that we will meet later. Most students of human evolution expressly do not intend the word *hominid* to include any ancient chimpanzees or gorillas that might come their way (not that very many have so far).

The living primates are a complicated lot, but what matters most about them and distinguishes them from all other creatures is their braininess.

The Brains of Primates

As a general rule, very clever animals have correspondingly big brains, but it's not quite as simple as that. Within any one class of creatures, bigger animals have bigger brains. The biggest brain of all belongs to the sperm whale, at around twenty pounds (8 kilograms), and even cart horses have substantially bigger brains than human beings.

What counts more is not the absolute size of the brain but the size of the brain relative to the size of the body. And what really counts is the thickness of the neocortex, the layer of tissue that lies on top of the cerebrum and is most associated with all those qualities that in humans we call creativity, thought, imagination, and so on. In relative brain size, mammals (and birds) score high compared with reptiles, amphibians, and fish, while primates as a group score high compared with other mammals. The brain of a monkey would be about thirty times bigger than the brain of a reptile of the same body weight, and about three times bigger than that of a similar-size dog. The brains of modern humans (and Neanderthals) are about three times bigger than an ape's of comparable body weight.

It's good to be brainy. Clever animals can outsmart animals that are less bright, and it seems that evolution has favored braininess. If we look at the fossil skulls of ancient mammals, whether of primates or dogs or rodents or pigs, in general the relative brain size has increased over the past few tens of millions of years. This means that different groups of mammals have *independently* acquired the quality of braininess and, presumably, the extra intelligence that goes with it. If big brains weren't worth having, natural selection would not have favored them so consistently.

But unless the lifestyle of an animal enables it to take advantage of its own cleverness, a big brain would not be a plus. To put the matter in context: if a catfish were to wake up one morning and find that it had the brain of Albert Einstein, it would do it no good at all. Its blinding scientific prowess would be totally wasted on its fellow fish, and its prodigious brain would require more energy every day than the pond could possibly provide, even if the catfish searched for food night and day. For such an animal in such a place, a big brain would be an encumbrance—a feature to be stamped out immediately by natural selection.

Big brains can evolve only in very particular animals that lead a very special kind of life—a life that enables them to feed well and that rewards any increase in brainpower. Clearly, many mammals and birds have been able to meet these requirements up to a point, but primates have done this better than anyone. So what is so special about primates?

Primates in general have sought out the high-energy, fatty-acid-rich diets that brains require. Some primates, to be sure, eat mainly leaves, like the leaf-eating monkeys of both the New World and the Old. Peculiarly, too, gelada baboons live largely on grass. But most primates live mostly on fruit, and virtually all of them eat some animal fare when they can get it, which boosts the diet in energy, protein, and yet more of the essential fats. Virtually all of the prosimians eat at least some insects—and some are mainly insectivorous and others will even catch and eat small vertebrates. Many small prosimians and monkeys also boost their energy by chiseling holes in the bark of trees with their pointed incisors and licking out the sap and gum.

Primates in general have the right kind of diet to promote braininess, but what is the payoff? What is the immediate reward for cleverness that makes having a big brain worthwhile?

There are two main views on this. The first—the traditional one—says that primates have become as brainy as they have because their brains operate in concert with their hands, which gives them a tremendous advantage. The second says that they have become as bright as they have because they are so social—and sociality pays. The traditional brain-hand idea has gone somewhat out of fashion, but there is serious mileage in it. Taken alone, explanations based on sociality just won't do.

The hand-brain idea begins with the hand. Arboreal creatures don't really need to grip the branches in which they spend their lives. Squirrels, for example, are wonderfully at home in the trees just clinging to the bark with their claws. Most bears are good climbers, considering their bulk, and though they grip the trunks of trees, they do so with their whole arms and legs. They cannot hold on with their hands alone. Primates are the only creatures of the trees that do this.

The gripping hands of some primates are located at the

ends of arms that are almost as flexible—thanks to the cunning engineering of shoulders, elbows, and wrists—as the tentacles of an octopus; they are able to move through almost all the angles of an entire sphere. No other mammal can come close to doing this. Because primates' arms are so flexible, they can easily use both hands for any task that cannot be done with one—like grabbing a big piece of fruit or threading a needle. Then again, because a primate takes its life literally in its hands as it leaps from branch to branch (they don't all leap, but many of them do), they need their brains for perfect coordination of hand and eye. Clumsiness is not an option.

In animals with such a way of life, big brains clearly do pay off. The more dexterous and agile the animal becomes, the better it is able to find nourishment for the brain. Whether it uses its brain for particular manual skills or just in the interests of agility, the brainier the animal is, the better its ability to make use of its difficult and complex environment; the more it can get around in the trees; and the more easily it can tease out insects from the bark or reach for fruits that for other creatures (apart from birds and fruit bats) would be out of reach. So there is positive feedback: big brains make better use of cunning hands and mobile arms; and the more cunning the hands become, the better they are able to feed the brain. So the two have coevolved, each egging the other on. In the making of tools, which chimps do and human beings do even more, the eye-brain interaction and interdependence is even more obvious. But all this depends on circumstance and opportunity. Pigs are bright, but they are stuck with hoofs. A highly imaginative pig would doubtless derive more benefit than a supremely gifted catfish could, but not enough to justify the enormous physiological strain of an overlarge brain.

The view that big brain size is based on sociality also has a great deal to recommend it. In fact, the two explanations work best when placed side by side, complementing each other. Primates, the brainiest creatures of all, are the most dexterous and among the best coordinated *and* they are among the most social.

There are some disadvantages in being social, which is why a lot of animals prefer to live on their own. Animals that live in groups have to compete for food, which they don't have to do if they first carve out a feeding territory and guard it. Also, animals in big herds attract the attention of predators more easily than they would if they moved around quietly on their own and kept their heads down.

In practice, the term *sociality* has many meanings. It can imply a simple tendency to cluster together, as emus or wildebeests do when migrating, or starlings do when they come home to roost. The different individuals within a flock or a herd almost certainly interact with one another far more than we can easily observe, yet we could not call a herd of migrating wildebeests a society. It is more like a football crowd: lots of individuals moving along together because they all want to go to the same place at the same time.

But there are many advantages to group living too, which explains why so many animals are social at least for part of the year. Many animals derive more direct benefit from living in groups, even if they are not interacting to any great extent. Vultures often wheel around in the sky together, each one keeping a sharp lookout and all of them cashing in when one of them spots a corpse. Many pairs of eyes are better than one pair, and if there is a dead zebra down there, that is far too much for one vulture to eat, so they might as well share it. Besides, when they peck away together, they tear up the spoils far more efficiently than one would on its own. Yet there is little reason to suppose that the different vultures in the flock know all the others as individuals. With many birds, even mixed flocks are common: different species operating side by side to their common benefit but going their own way when the day's work is done. Often, for example, in tropical forests, we find very mixed collections of birds following columns of predatory ants and picking up the many small animals that are panicked by their advance.

In true societies, though—as in prides of lions or packs of wolves or troops of primates—the individuals stay together over time. Each individual knows all the others personally—knows whom to defer to, whom it is safe to bully, and whom to trust. In general, if it pays to live in a society at all, then big societies can compete in the wide world more effectively than small societies. But in big societies, the social relations become very complicated indeed. As the number of individuals in the society doubles, so the number of possible one-to-one interactions between those individuals increases at least four times. Increase the number four times—say, from five individuals to twenty—and the possible number of one-to-one interactions increases nearly twenty times. If the different individuals start to subdivide into little gangs of friends and allies, then the number of possible interactions increases more or less indefinitely—and even more than is obvious, because alliances of animals, like alliances of governments, are all too likely to last only as long as it pays for the individuals not to betray their buddies, however fervently they swear their loyalty. The complexities of group living are horrendous, and only big-brained animals can manage it. But again, as with the hand-to-brain interaction, we find that brain size and the size and complexity of the society increase in concert. The bigger the brain, the bigger the possible society and the greater the ecological advantage. But at the same time, the more the society grows, the more natural selection favors big brains, because only the brainiest creatures can manage it. Hence, another positive-feedback loop, just what is needed for rapid evolution.

In practice, the social arrangements of primates are just as various as every other feature of primates; and primate societies, at least in many ways, are the most complex of all. Human beings demonstrate this complexity most spectacularly, but chimpanzees and gorillas and even some monkeys, when we look closely, are not so far behind.

The Social Life of Primates

Everything that animals do is related to their bodily form—their size, their shape, the proportion of their limbs—and everything that applies to them also applies to us. For instance, we might suppose that body size has nothing much directly to do with social life. Human folklore contains many a reference to pixies, leprechauns, and the like, and we tend to take it for granted that if the little people had a family life at all, it would be just like ours: Mommy, Daddy, with 2.3 even littler people who take the same amount of time as our children do to grow up and find an honest job appropriate to their kind—upsetting applecarts and burying pots of gold and all the other things that little people feel obliged to do.

But it doesn't work like that. Body size, which is largely related to diet, which is largely related to place and opportunity, profoundly influences the overall pattern of life, which affects mating strategies, which in turn affects social structure. And social life affects the evolution and development of intelligence and all that goes with it, and this, in turn, influences the social life. Since primates have such a huge range of body sizes and diets, they illustrate these principles beautifully.

In general, big animals take longer to grow to full size and hence to reach sexual maturity than small ones do, basically for reasons of simple physics. Animals with big brains tend to take the longest of all, perhaps because their brains require so much energy to maintain, and even more for growth. Once they are mature, big animals tend to live longer—perhaps because, once they achieve maturity, they are less vulnerable than small animals and *can* live longer—and they are adapted to take advantage of their long life. So we find that mouse lemurs can breed when they are just one year old, give birth after a two-month gestation, and then produce more offspring roughly at ten-month intervals. But they live more than fifteen years only if they are lucky.

Gorillas, on the other hand, are not ready to breed until they are ten years old, then have a nine-month gestation, and then can give birth every four years or so. They sometimes live until they are forty.

Humans living in hunter-gatherer societies tend to show much the same general pattern as gorillas, but at a slower pace. At least in bushman societies, the girls may reach menarche at around age thirteen, but they are not physically big enough and strong enough to have babies until they are about nineteen. From then on, they typically have one child about every five years, with the last one born when they are in their late thirties—not more than an average of five in all. If women in modern societies each have five children, then the population booms. But out in the wild, where the hazards are great, five is roughly the number needed to keep the population stable.

In all species, both males and females need to reproduce, whether together or independently, depending on the species (if they don't, their own particular lineage dies out). But their roles in reproduction are different, and this is especially so in mammals. To a large extent, the male's job ends once he has made the female pregnant, although a few male mammals do hang around and help, sometimes a little and sometimes a lot, to ensure that the resultant infant survives. Female mammals, however, not only have to get pregnant, they also have to look after the babies. Some female birds get out of this and leave the child care to the males, but for female mammals, that is not an option: baby mammals need milk, which only mothers can supply.

This built-in asymmetry in male and female roles spills over into sexual politics. Some primates, such as gibbons, form faithful, monogamous pair-bonds. In monogamous species, the males and females tend to be much the same size, because, after all, both share the chores, and with luck all the males as well as all the females in any one generation finish up with partners and produce offspring. But many male primates are polygynous, having more than one "wife" at the same time—such as chimps, gorillas, and many monkeys, including baboons. In polygynous societies, all the fertile females get to breed, but most of the males do not. If a male in a polygynous society is to breed at all, he has to chase away all his rivals, which means he has to be big and strong. So in polygynous species, we see marked sexual dimorphism; the males are much bigger than the females and often have huge and very dangerous canine teeth. The canine teeth of a large baboon can be bigger than a wolf's, partly for show but also a serious threat.

There are further complications in some primates' social lives. Hanuman langurs, big monkeys from Asia, sometimes live in groups with only one adult male, several females, and offspring of various ages, including small infants that are still being nursed. Every now and again, the resident male will die or, before this happens, be thrown out by some incoming gang of bachelors—bachelors who left their own troops when they reached puberty and then teamed up with their fellow exiles. One of the male newcomers then takes over as the boss.

Sometimes the new boss then embarks on a truly chilling course of action. He starts killing the infants that were sired by the previous boss and are still feeding from their mothers. This was first observed in Hanuman langurs in the 1960s, and infanticide has been observed in several primates since, including numerous species of lemurs, howler monkeys, leaf monkeys, guenons, Savanna baboons, Chacma baboons, chimpanzees, and even mountain gorillas (famed, after the studies in particular by Dian Fossey and their tête-à-tête with David Attenborough on British television, for their gentleness). It isn't only the male primates who kill infants, but they are the usual culprits. Infanticide has not been seen often, but in red howler monkeys and mountain gorillas, it is said that about one in eight of the infants is killed by an adult male, which thus accounts for about a third of all infant deaths.

Masses of scientific data and a lot of intriguing anecdotes show how very subtle the social interactions of primates have become. In practice, the big alpha males that apparently are in sole command don't necessarily have it their own way. One pleasant illustration comes from the Monkey Sanctuary in Looe, in Cornwall, which has a thriving colony of woolly monkeys. A few years ago, the dominant male of the day finally gave up the ghost, leaving the succession to two young pretenders. One of them was big, strong, and cocky, and, as expected, he assumed command as if by right. The other was a quieter individual, more dignified. Woolly monkey societies are not crude. In woollies, all the adults, including the males, are expected to take good care of the infants, and the infants know this and sometimes take liberties. For a time, all went fairly well down in Looe with the new leader in place. Then one day, the new boss lost his temper with an infant that jumped on his back, and he knocked it away with some violence. Nothing much happened for a day or two, but then the females ganged up on him and made it perfectly clear that he no longer had any authority in the group. The quieter of the two males, an altogether more balanced individual, took over.

Alliances—friendships—of many kinds are common within primate groups, and these friendships may endure for life. In all contexts—not just for the protection of infants—individuals with allies generally fare much better than those without. As the expression has it, it's not what you know, it's who you know. Regarding baboon societies, there are many tales of young males cooperating in mating. One of them undertakes to distract the boss male while his friend has his way with one of the females. Later the favor may be returned.

At Howletts Wild Animal Park in Kent, Dr. Jennifer Scott found among gorillas examples of extreme deviousness of the kind that has been called Machiavellian intelligence. For example, a low-ranking female who wanted to get back at a high-ranking female who had been giving her a bad time would suddenly run away from the high-ranking female, yelling and screaming and pretending to be hurt. This would attract the attention of the alpha male, who would then assume that the high-ranker must have been bashing the low-ranker and would then in turn punish the high-ranker. (Alpha males in gorilla society act as both judge and chief of police.) Small children have been known to do exactly the same kind of thing to get their elder siblings into trouble. On the other hand, if the alpha male at Howletts dished out punishment unjustly, the females would gang up on him—perhaps not immediately but within a day or two—just as the female woolly monkeys did at the Monkey Sanctuary. They would not physically attack the alpha male (which would be very foolish, and somewhat sacrilegious), but they would make it clear that they were no longer "talking to him." In the most sophisticated primate circles—both among monkeys and great apes—the alpha males emphatically do not rule by might alone. They need approval and consensus; and they go to considerable lengths to ensure that they have it. By such means, the simple biological imperative to murder offspring that may have been sired by rival males is much mitigated.

In general, the social lives of prosimians are not so intricate, but neither are they based on a straightforward dominance of strong over weak, or male over female. The pioneer of lemur watching was Alison Jolly, now at the University of Winchester. In the 1960s she was surprised to see a senior-ranking female ring-tailed lemur calmly take a choice piece of food from the dominant male, who walked away equally calmly and foraged somewhere else. As far as he was concerned, this was par for the course. This was a female-dominated society and he knew his place.

So, what has all this got to do with Ida? Everything, is the answer. We cannot understand what we do not know except by comparing the unknown with the things we do know. By looking at her bones in a nearly complete skeleton, we can see where she fits among the panoply of primates, that is, who her closest relatives are—whether she is more anthropoid or prosimian. From her size and the general shape and length of her limbs, and whether her hind limbs are longer or shorter than her forelimbs or of the same length, we can judge how she moved. From all this, by extrapolation, we can make a fair guess at the kind of society she kept—whether, if she had lived to be an adult, she was destined to be the lifelong partner of a faithful "husband" or a member of a harem dominated by some erratic male. We cannot know any of this for sure, of course, but we can make good, informed guesses. And the flow of information works both ways: once we get a handle on who she was, we will have a much clearer idea of primate evolution as a whole—which is our own.

CHAPTER SIX

PRIMATE EVOLUTION

Few scientists are as restless as paleontologists, who search for and study the fossils pulled from the earth at Messel and from other sites around the world. And no subject is more tantalizing. In two hundred years of intrepid searching, scientists and amateurs have revealed thousands of creatures that are no longer with us, and not just individual species but entire groups: genera, families, orders, and classes that had their hour upon the stage and now are gone forever. Taken all together, these fossils provide at least a broad-brush picture of life on Earth over the past 2 billion years.

In framing his theory of evolution, Charles Darwin did not actually make much use of the fossil record. But if there hadn't been any fossils to look at, neither he nor anyone else would ever have supposed that living creatures evolved from creatures that were often quite different from them and that lived deep in the past. And no one could ever have realized just how long it took those ancient animals and plants to attain their modern forms.

The fossil record as a whole is one of the world's wonders. Some religious fundamentalists still doubt whether evolution happened at all, but it might indeed be argued that the fossil record itself is a gift: a fragmentary but generous insight into the past, of which human history is a small but significant part.

And yet, when it comes to the fine details—what exactly these ancient creatures were, how they relate to modern types, and how they really lived—the record frustrates. Sometimes entire continents where we know important things must have happened seem totally free of fossils from the critical periods; the rock that may once have contained the vital traces has simply eroded away. More often than not—especially when we are dealing with primates—we have nothing to go on but a fragment of lower jaw (the bone of the mandible is particularly hard and prone to fossilize) or a few teeth (which rot readily enough in life but last very well after death).

Primates in general and Ida in particular illustrate both the joys and the frustrations of the record. One thing is clear: the number of primates that lived in the past and are extinct far outweigh the number that are living now, even though there are hundreds of living types. On statistical grounds alone we know that this must be true—the fossil record tells us that most species of mammals survived on Earth for only about a million years before going extinct or evolving into something recognizably different—and as we will see, primates have been around for many millions of years, so there has been ample time for many complete changes of personnel.

Some animals fossilize far better than primates do, and from them we can see directly how enormously past species as a whole may outstrip the present. Today, for example, there are only two living species of elephant, one in Asia and one in Africa. Yet at least 150 species of elephants and elephant relatives are known from the past 50 million years or so. Now the world has only five species of rhinoceros—two in Africa and three in Asia—but around two hundred rhinos and close rhino relatives are known from the past (and the rhino group as a whole probably arose not in Africa or Asia but in North America, where the last ones died out around 5 million years ago).

Though primates do not fossilize well, we still know of many more genera from the past than from the present, and most of those genera would have contained several (even many) different species. Our own genus, *Homo,* evolved relatively recently by geological standards, but we know of at least half a dozen species from the past, and some paleontologists suggest that in that brief time, there were twenty or more. However many ancient species or genera we know about, we can be certain that there must have been many more—perhaps many times as many—of which we have no inkling at all and perhaps never will because if they did form fossils, those fossils have long since eroded away.

How Old Are the Primates?

Though the record isn't as complete as we'd like, we can still guess that primates are far older than might be expected. If we ask how long human beings (or chimps or spider monkeys) have been evolving, the answer might be, around 3.8 billion years, the same as mushrooms and earthworms and oak trees. All creatures now on Earth descended from the same primordial ancestors, who probably lived around

3.8 billion years ago; and the ways in which our ancestors evolved *before* they became recognizable as people, or oak trees, or mushrooms, are relevant too. We all carry legacies from the very earliest times.

So the question, How old are the primates? should be taken to mean, What is the date of the first-ever primate?—the one from whom all living primates are descended. Or, as the paleontologists tend to put it, What is the date of the primates' most recent common ancestor? In truth, whatever way you frame this question, the answer is that nobody knows—and no one can ever know with absolute certainty. But various facts all point to roughly the same conclusion.

First, we can look at the DNA of living types. Evolution implies that creatures change over time, and this means that their genome (the sum total of their genes) changes over time too, because the form of the creature depends to a large extent on its genome. The DNA from which genes are made also changes randomly, in ways that do not necessarily affect the function of the creature. These small, ineffectual changes simply accumulate in the genome, and although they occur randomly, they tend to occur at a fairly steady pace. So the steady accumulation of minute but functionless changes provides us with a "molecular clock." We can guess when any two creatures last shared a common ancestor by measuring the difference in the functionless bits of their DNA.

Then we can look at the fossil record, which helps us set the molecular clock. For example, the difference in nonfunctional DNA between humans and chimps tells us that they last shared a common ancestor around 5 million years ago, which is so recent that it might seem surprising. But we've discovered fossils that date from about 5 million years ago and look like the kind of creature that a human-chimp common ancestor might have been. When the molecular clock tells us what age fossil to look for and a fossil turns up that is the right age and is in the right place, it suggests that the molecular clock was indeed accurate.

Fossils tell us that a particular creature did exist in a certain place at a certain time, but they can't tell us what other creatures might have existed in that place at some earlier time or in some other place. The chances of finding any fossil are remote. The chances of finding the very first individuals from an entire new lineage are remote beyond imagining. The chances diminish even further because the very first individuals of a new lineage must have been rare to begin with. And then, even if by some miracle of miracles we did find the first-ever common ancestor we were looking for, we would not know for certain that that was what it was. Fossils do not come gift-wrapped with labels attached.

But from the limited amount we do know, we can infer a great deal. First, by the time we get to the Eocene—Ida's period—prosimian primates such as lemurs and tarsiers were already common and widespread throughout the landmasses that now form Eurasia and North America. There is plenty of reason to think that the very first primates arose in Africa and that we haven't found the earliest ones from Africa simply because there aren't any good fossil sites from the right period. From this we can guess that in the case of the American and European fossils, we are really looking at the margins. In addition, although the Eocene fossil primates are all prosimians, they are already immensely varied. Thus, we can infer that they had already been evolving, and spreading themselves out, over a very long period. That is, they must have arisen a very long time before.

There are primate fossils from the period before the Eocene—the Paleocene—but even in those early times, they were already very varied and widespread. This means that for the very first stirrings in the primate line, we have to look even farther back.

Before the Paleocene epoch was the Cretaceous period— and not only a different period but a different *era,* right back in the Mesozoic. The first primates almost certainly appeared at least 80 million years ago, and probably nearer 90 million years ago, and though we're taught that mammals didn't come into their own until after dinosaurs had gone, this takes us back to dinosaur times.

Mammals and Dinosaurs

Ideas about dinosaurs were first framed in Victorian times by British scientists, with significant input from America, and biology has been carrying that Victorian legacy ever since. For the British, the nineteenth century was the age of Empire. For the Americans, it was their first taste of economic and political might. Both nations were convinced that they owed their supremacy to the natural talent and merit of their people.

Science is supposed to be immune from emotional input, but the reality is very different. The theories of any given period reflect the zeitgeist, the spirit of the age; and the leading scientists of the nineteenth century grew up in societies that believed, absolutely, in their own innate superiority. Late nineteenth-century biologists sought to explain this superiority in biological terms. Since they were humans, and humans are so powerful, then humans must be superior to everything else. And since humans are mammals, it followed too that mammals must in general be superior to nonmammals. In truth, Darwin was far less wedded to this line of thought than many of his contemporaries and successors were. But his peers and disciples used his ideas to support their own preconceptions, interpreting natural selection to mean that the strong must bash the weak. They considered it obvious that Europeans and Americans were flourishing because they were indeed better than the rest; and they assumed, more broadly, that mammals were the dominant animals because they were better than other creatures. This too is how they read the evolutionary record.

The world's knowledge of extinct creatures grew dramatically in the nineteenth century, with huge expeditions to fossil sites that had sometimes remained undisturbed since they were first laid down. Many of the sites looked (and still look) irredeemably barren. But many a modern desert was once tropical forest or temperate plain, teeming with life. Particularly striking, at least from rocks of the era that came to be known as the Mesozoic, were the dinosaurs. Many were extremely large, far bigger than any other land-based creature, and relative to their enormous bulk, they had very small brains. And quite suddenly, at a date that we now know to have been around 65 million years ago, the dinosaurs all disappeared.

The record of mammals is the complete opposite to that of the dinosaurs. Victorian paleontologists found few signs of mammal remains in the Mesozoic rocks. But they did find them, in increasing number and with increasingly impressive variety and bulk, after the dinosaurs disappeared. In short, the fossil record showed a sudden shift from a world dominated by big reptiles—not just dinosaurs but also plesiosaurs and ichthyosaurs and the giant lizards known as mososaurs in the seas, and the pterosaurs in the skies—to one dominated by mammals.

Many theologians—and the scientists of the early nineteenth century, whose education for the most part was steeped in theology—saw in this about-face the hand of God. Clearly He had decided that the dinosaurs had had their day and He decided to replace them with something different. But by the late nineteenth century, this kind of thinking was losing its grip. By then, most scientists sought to explain everything they saw in rational terms: cause and effect. Natural selection—at least when interpreted as "the strong bash the weak"—seemed to provide precisely the explanation they needed. Dinosaurs disappeared because by the end of the Mesozoic, a few mammals were already coming onto the scene; and because they were mammals, they were obviously superior; and being superior, they simply ousted the dinosaurs, which were too big and stupid for their own good. None of the fossil mammals that were found in Mesozoic rocks were bigger than a badger, and most were smaller than rats, and none had big brains. But by their superior wit and their general ability to resist changing conditions, they simply pushed the dinosaurs aside. Or so the theory had it. I have read in more than one place that the first mammals simply ate the dinosaur eggs, since it was taken for granted that nothing as brutish as a dinosaur could possibly indulge in parental care.

The fossil record and science in general have moved on

a great deal since the nineteenth century, and, astonishing though it would seem to anyone with a traditional zoological background, it's now clear that mammals are extraordinarily ancient. The very first identifiable mammals are at least as old as the very first recognizable dinosaurs. Both date from over 200 million years ago. The mammals are not so prominent in the fossil record, but that is not because they weren't there but rather because as long as the dinosaurs were around, the mammals were confined to the ecological margins; hole-in-the-corner creatures, they scuttled around on the forest floor, living largely on insects, ecologically equivalent to modern-day shrews. Far from outwitting and outgunning the dinosaurs, they were forced into near insignificance for as long as the dinosaurs were around, which was more than 130 million years. Birds came on the scene much later than mammals—not until about 140 million years ago, in the mid-Jurassic. By Cretaceous times, birds were already very varied and ecologically significant. But then, birds descended from dinosaurs—which means in effect that they *are* dinosaurs. They were part of the general dinosaur dominance (and to a significant extent, they still are).

What killed off the terrestrial dinosaurs was actually an asteroid. Luis Alvarez first suggested this informally in 1980, and radical though the notion then seemed, it is now the orthodoxy. The asteroid that did the job would not have been particularly big—perhaps about six miles (10 kilometers) in diameter. But physical theory shows that if such a bolide struck the Earth at, say, around sixty-two hundred miles (10,000 kilometers) an hour (a modest speed by astronomical standards), it would send up a cloud of debris into the stratosphere that would blot out the sun's rays and plunge the Earth into an icebox phase—long enough to eliminate all creatures that were not able to adjust. Since it is clear that some dinosaurs survived some seriously cold episodes during the Mesozoic, it is not obvious why they could not resist this particular setback. Nonetheless, the evidence suggests that such an asteroid did strike the world about 65 million years ago, and that the climate did change afterward, and that this is what finally did in the big reptiles.

The mammals pulled through *because* they were hole-in-the-corner creatures, and because they were warm-blooded—able to generate their own body heat and stay at a steady temperature independent of their surroundings. This scenario radically changed the conventional view of dinosaurs and mammals, and of their relative abilities. Many biologists now point out that the dinosaurs taken together were the most successful land animals of all time—certainly the most successful big ones. They dominated the Earth for 130 million years. Neither were they dolts. The distribution of their fossil bones suggests that many lived in complex social groups and that some at least were good mothers. It seems too that far from ousting the dinosaurs, the mammals stayed in their shadow through all that long time. They came into their own only after some freak astronomical event changed the world in a way that happened to favor them. They stepped into the breach, filling the vacuum that the dinosaurs left in their wake.

Even when the dinosaurs were still around, mammals made some significant strides in the Cretaceous. The two most prominent groups of modern mammals, the placentals (such as the primates) and the marsupials (such as the kangaroos and koalas) clearly shared a common ancestor, and it seems that those two groups first split apart to form separate lineages about 130 million years ago, early in the Cretaceous. Most of the mammals we do know about from the late Cretaceous were still very primitive and "generalized" (most still look like shrews), but even so, among them we can perhaps see the first traces of modernity — the first more or less recognizable rodents and so on.

This leads us to one last irony. Not so long ago, many people seemed to have it in their heads that ancient human beings, peremptorily known as "cavemen," shared their humble dwellings with dinosaurs — as immortalized not only by the wonderfully anachronistic Raquel Welch but also by the wonderfully anachronistic *Flintstones*. Now, of course, everyone knows that *The Flintstones* is a fantasy, and we repeat the accepted wisdom that human beings and dinosaurs missed each other by more than 60 million years.

But the pendulum has swung too far. Our ancestry did begin before the first cavemen, whoever they were. Human beings have inherited from our nonhuman primate ancestors a lot of the features we consider characteristically human. The very first of those primate ancestors shared the world with some spectacular dinosaurs, including *Triceratops,* which filled the niche of the large pillar-legged herbivore now occupied by rhinos; the swift, bipedal, carnivorous velociraptors immortalized by Steven Spielberg in *Jurassic Park* (although many of the park's incumbents were Cretaceous); and, most famously, the engagingly nightmarish *Tyrannosaurus rex*. So our lineage did interact directly with the dinosaurs, and since every creature must adapt to the presence of all the rest, we can reasonably conclude that dinosaurs influenced the course of primate evolution. *The Flintstones* is still a fantasy, but it is not quite as fantastic as we have all been brought up to believe.

What Do We Really Know of Ancient Primates?

At first sight, the primate fossil record is quite encouraging. We have some significant fossils, and the oldest of them do date back almost to the K-T boundary. It is even suggested that some known Cretaceous mammal fossils were early primates.

Yet if we lay out all the fossils that are known and try to construct a family tree from them, the result is, well, sketchy. Evolutionary paths do not run smoothly, and they are certainly not straight. In the first flurry of post-Darwinian zeal, many biologists hoped and expected to find successions of fossils inexorably mutating in an undeviating line from an obviously primitive ancestral state to modernity. In most cases, however, the record is far from complete. When long and more or less continuous records do exist, they sometimes show lengthy periods of stasis, when the ancient creatures did not apparently change much over many millions of years, and indicate no progression at all (this is particularly true of many marine creatures, such as clams, which may seem virtually unchanged over hundreds of millions of years).

The fossil records of the mammals and other animals that are with us now—at least those that are reasonably well known—invariably show all kinds of deviations. Horses, for example, have taken many forms over the past 50 million or so years—some big, some small, some grazers, some browsers, some with one toe, and some with three. It is good to be big if you live on grass, so that you can cover large distances and subsist on low-quality food, and since it's a good idea for mechanical reasons to have just one toe, and one with a hard tip to it, we can reasonably assert that if grassland is abundant, then big, one-toed horses are likely to be favored—and we shouldn't be surprised if such creatures do indeed emerge in the record. But we cannot say that grassland was bound to become as common as in fact it has, and we cannot be sure that the animals we think should do well will in fact do so. The most realistic expectation, in short, is not that the evolutionary tree should proceed in a straight line, but that it should be extremely jagged.

The best we can ever say is that a particular fossil probably comes from the kind of animal we would expect a given creature's ancestors to have been like; and that the animal whose fossil bones we have is probably related to the ancestor of some modern type. Of course, we might find the direct ancestors we are looking for, but we are far more likely to find members of some small side branch, which probably was on an evolutionary path that has left no modern descendants. This is just as true of primates as of every other creature. Indeed, as we will see, it came as a great shock in the middle decades of the twentieth century to discover that the evolutionary tree of human beings is just as bushy as that of any other creature.

If paleontologists have no idea at all of what they should be looking for—no expectations and no preconceptions—then they will not recognize the fossils they are looking for, even if they find them. On the other hand, if their preconceptions are too rigid, then they will miss the significance of anything surprising. So, modern paleontologists consider the bushy primate tree and visualize the possibilities.

It is a reasonable working hypothesis that the very first primates resembled modern tree shrews, of which there are eighteen known species in six genera, and which live in various countries and quite a few islands in Southeast Asia and parts of India. They were first called tree shrews at the end of the eighteenth century by William Ellis on his travels with Captain James Cook. It's an obvious enough name for an animal that lives mainly in tropical forests and roughly resembles both squirrels and shrews, but tree shrews in truth are not particularly arboreal (they spend much or most of their time on the ground), and they are not related to shrews. Like shrews, they do eat a lot of insects, but unlike shrews, they also eat seeds and nuts.

In general shape and size, tree shrews resemble several of the many kinds of mammals that lived in the Mesozoic, and—especially significant—in various anatomical details they also resemble primates. For a time, indeed, tree shrews were classified as primates. Now they are given their own order—the Scandentia—but they are thought to have shared a common ancestor with primates sometime in the Cretaceous. But tree shrews, more than primates, have retained the ancestral features, at least in a general way. All in all, then, they provide a good search image for paleontologists. They are the kind of beast we probably should be looking for.

It isn't just a question of search image, however. We know from Darwin that all living creatures must have ancestors; and the fossil record from any one time in the past could show traces of those ancestors. Thus, we can reasonably expect that the earliest fossils of the Tertiary would include early primates, or the precursors of primates. It could of course be that all the primate fossils have been obliterated, but there is

Atractosteus strausi, an early fish from the Messel Pit.

Europe's second-largest fossil fair, in Hamburg, Germany. It was here in December 2006 that Jørn Hurum was first shown a color photograph of Ida.

Dr. Jørn Hurum, the scientist who uncovered Ida.

Dr. Holly Smith at the Duke Lemur Center, North Carolina.

(Left to right) Dr. Jørn Hurum, Professor Philip Gingerich, and Dr. Jens Franzen at the Natural History Museum, University of Oslo.

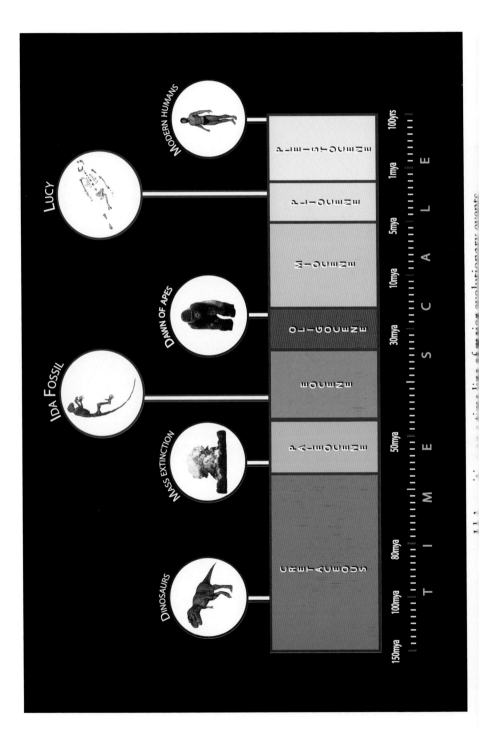

A representative timeline of major evolutionary events.

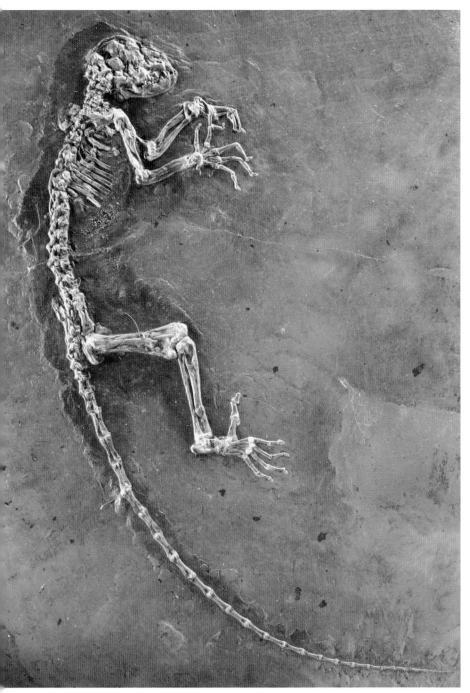
Ida fossil in the matrix.

A full shot of Ida in contrast.

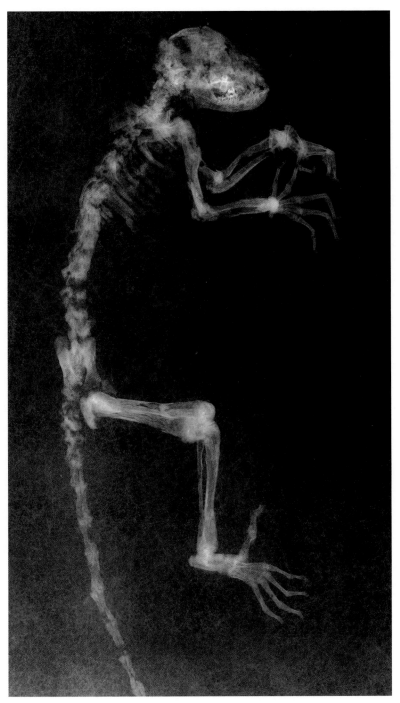

A full-body X-ray of Ida, conducted by Jörg Habersetzer at the Senckenberg Research Institute, Frankfurt, Germany.

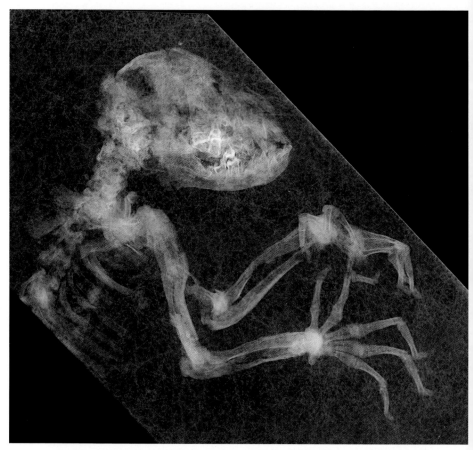

An X-ray showing Ida's hands, conducted by Jörg Habersetzer at the Senckenberg Research Institute.

Ida's hands. The rounded fingertips are typical of nail-bearing fingers.

A closer look at Ida's grasping hands.

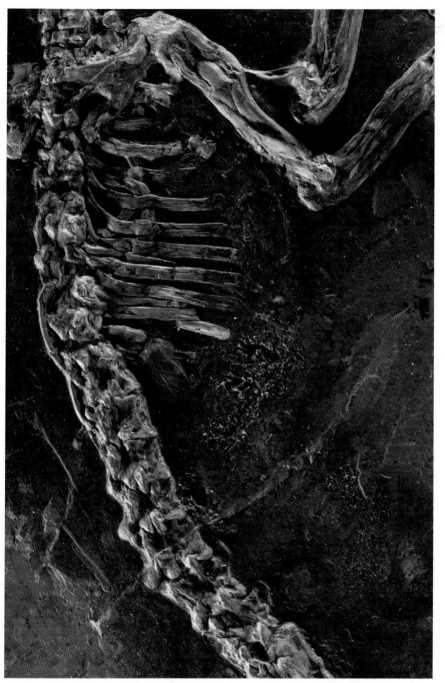

The content of Ida's stomach is incredibly well preserved.

A closer look at Ida's gut content.

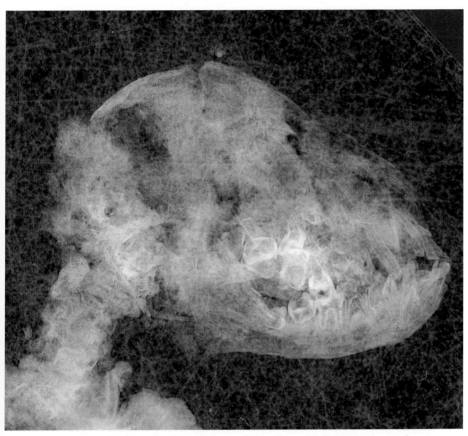

An X-ray of Ida's skull, conducted by Jörg Habersetzer at the Senckenberg Research Institute. The X-ray reveals that Ida has well-formed adult teeth behind her baby teeth.

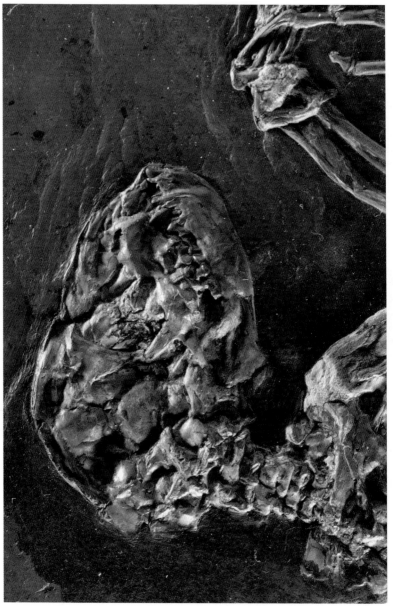

Ida's skull.

A close-up of Ida's skull, showing that her canine teeth are still milk teeth, though X-rays show the permanent teeth behind them. The teeth reveal that Ida was less than one year old when she died. Her teeth are a combination of broad-basined teeth and incisor teeth with slicing edges—typical of a fruit-and-leaf eater.

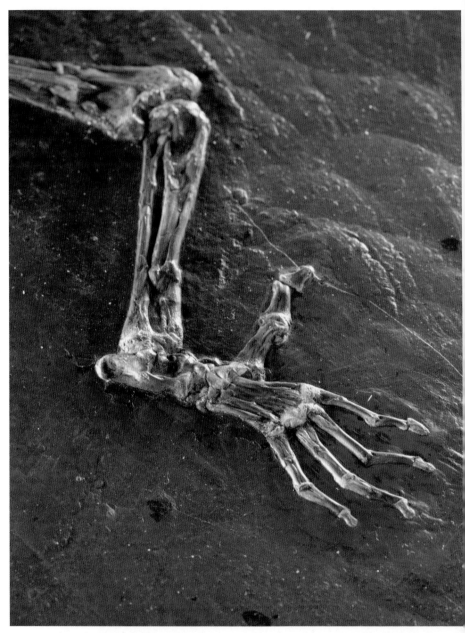

Ida's foot, lacking the grooming claw on the second digit that is typical of all lemur primates.

no good reason to assume this. So, in addition to the search image, we also bring logic to bear and ask, Of all the fossils that we know about from the early Tertiary, are there any that could plausibly be on the primate lineage?

Are there any fossils from the late Cretaceous or the early Tertiary that roughly resemble modern tree shrews but are not already too specialized to be primate ancestors? The answer is that there are. Plenty. They are formally known as Plesiadapiformes—roughly meaning "primitive creatures that somewhat resemble *Adipus*" (*Adipus* being the first bona fide primate ever to be identified). Less formally they are called archaic primates or, sometimes, proto-primates.

The Archaics

The archaic primates were not big. Indeed, the first two species to be formally described—wonderfully called *Purgatorius*— were about the size of a mouse. Their cheek teeth had sharp cusps, like those of present-day mouse lemurs and the dwarf bush baby, and they too must have been insect eaters. Their fossils come from eastern Montana, and they date from the early Paleocene and the late Cretaceous. The most familiar and widespread of the archaics is *Plesiadapis* from the Paleocene, known from both Colorado and from France. *Plesiadapis* was first described and named in 1870.

A wide variety of archaics are known both from North America and from Europe, and since all archaic primates have been found in the Northern Hemisphere, it's a reasonable bet that this is where they originated. In other words, primates as a whole probably arose either in North America or in Europe. In later times, most of the really big innovations in primate evolution (including human evolution) took place in Africa, and that has prompted some to suggest that Africa might have produced the very first archaics. The question arises, then, Why aren't any archaics known from Africa? The answer is simple: Africa has very few rocks that date from the early Tertiary. Very few African mammals of any kind are known until the Oligocene, the epoch after the Eocene, by which time there were obviously a great many. So, though the present evidence suggests that primates may be American or European, the present evidence may be seriously misleading.

The fossils of archaics, and of many later creatures, also seem to have a very odd distribution. They have been found in Belgium, Germany, and France; and very similar beasts were found in Montana and Wyoming. The reason for this distribution is continental drift. In the late Cretaceous and the Paleocene, western Europe and North America formed one continuous landmass; the present-day Faroe Islands, Iceland, and Greenland were all connected, and they bridged what is now the North Atlantic. To be sure, Wyoming and France were still a long ways apart, but we should keep in mind that the land in between would have been perfectly hospitable and the animals were around for millions of years. Even if they had spread themselves out by only a few yards a year, they would have had ample time to cover such distances.

By the mid-Paleocene, around 60 million years ago, the archaics had diversified within the North American–Eurasian landmass to form at least four distinct families—and there could well have been many more, because we surely don't know of all of them. Most of them were still the size of a mouse or a rat—beyond doubt the primates started small— but a few had reached the size of a domestic cat. As time passed, and especially within the bigger types, their back teeth became less and less needle-like—less and less specialized for eating insects—and more and more adapted for fruit and leaves. In short, their diets became more like that of most modern primates.

Even so, the archaic primates were far from modern. Their skulls lacked the bony strut behind the eye (the postorbital strut) that is characteristic of modern primates. Their brains were very small and primitive. Their thumbs and big toes were not opposable. Fairly big chunks of *Plesiadapis* skeletons have been found here and there, and they tell us that this star archaic had shortish, stocky limbs with fingers and toes that ended in claws. The claws were probably for climbing, but even so, the archaics as a group do not seem to have been particularly well adapted for a life in the trees. Like the ill-named tree shrews, they probably spent a great deal of time on the ground.

Widespread and successful as the archaics clearly were in the Paleocene, by about 50 million years ago, in the early Eocene, they had begun to decline. The record suggests that by the end of the Eocene, there were none left in Europe and only one in North America. Ecologically it seems they were replaced by the rodents, which radiated enormously, giving rise to a great many more species. Increasingly too they were replaced by prosimians: the first of the true primates.

The First of the Prosimians

Some of the earliest-known fossils of true primates come from Wyoming, and some come from Belgium, France, and England. Again we see the North American–European continuum, and they date from the early Eocene. One of them is called *Teilhardina*, named after the great Jesuit paleontologist Pierre Teilhard de Chardin, and it is, again, the size of a mouse; the other is *Cantius*, and it is the size of a rat. Each of them has been found both in North America and in Europe.

Teilhardina and *Cantius* are so different from each other that they are placed in different zoological families—*Teilhardina* in the Omomyidae, and *Cantius* in the Adapidae. About forty genera of these early prosimians are known, and they all belong to one or the other of these two families. Since the two oldest known had already diverged to form two different families, it's plain to see that their common ancestor must have lived a long time earlier—way back in the Paleocene, and perhaps even before.

In contrast to the archaic primates, these Eocene prosimians clearly were adapted to life in the trees. This is evident from their limbs and hands and the opposable big toes on their feet. In both families, many of them had long hind limbs typical of clingers and leapers, and some had extra-long ankle bones, increasing the spring, as in a modern tarsier. The pattern of cusps on their cheek teeth suggests they ate fruits and leaves. Their eyes pointed forward for stereoscopic vision, and some had very big eyes, suggesting they were active at night, like tarsiers and bush babies. Their brains were big too by the standards of small Eocene mammals.

But it isn't exactly clear how the omomyids (members of the Omomyidae) or adapids (members of the Adapidae) relate to present-day primates. They were of prosimian grade, with true primate features such as the complete bony orbits, and their teeth and skeletons have much in common with those of modern prosimians. Not surprising, neither of them has any of the special features of the anthropoids—the apes that most closely resemble humans. It has often been suggested that the omomyids gave rise to the modern tarsiers, and the adapids (which in general were bigger than the omomyids) gave rise to the lemurs. But this is far from certain. It is even possible that neither of them is ancestral to any of the modern types—both of them might simply be side branches—but that is perhaps being a little too pessimistic. Neither is it clear how the adapids and the omomyids relate to the anthropoids, if at all.

It is reasonable to suggest, though, that if they weren't the ancestors of the anthropoids, then something very like them surely was. So who were the world's first anthropoids?

The World's First Anthropoids

It would be nice to be able to provide details on the emergence and spread of the anthropoids, describing the first monkeys of the Old World, the first anthropoid apes, and the first New World monkeys, and explaining how they all became the kinds of creatures we see today and ended up living where they do. It would be nice, but it is not possible. Too many fossils from critical periods are missing, including, perhaps of greatest importance, most of the Eocene from Africa. As always, much of our knowledge is based on the fossils from just one particularly favored site. For anthropoid origins, we owe a great deal to a spot in Egypt about thirty-seven miles (60 kilometers) south of Cairo that dates from the late Eocene (around 40 million years ago) through the mid-Oligocene (to about 31 million years ago). This site, called the Fayum Depression, is just another stretch of rocky, forbidding desert. But it once was tropical forest. Many of its fossils are beautifully preserved, including, among many others, the needle-thin foot bones of a jacana—also called a lily-trotter, which shows there was a lake there once, complete with lilies.

Thanks largely to the tireless efforts of the American paleontologist Elwyn Simons, who began work there in the 1960s, the Fayum has also yielded an impressive array of primates—perhaps as many as seventeen different species. These include an omomyid; prosimians, some that are like modern tarsiers and others resembling modern lemurs; and some of the earliest-known bona fide anthropoids. The latter are already various enough to suggest that the very first anthropoids must have arisen in the Eocene. All this, as we will see, enriches the story of Ida.

Mostly based on common sense, and from our knowledge of modern primates, we can sketch out, at least roughly, the probable pattern of anthropoid evolution. Presumably in the beginning—probably in Eocene Africa—there was just one basic anthropoid lineage. At a very early stage, some say, this line split to form the line that gave rise to the platyrhines (the New World monkeys) and the catarrhines (the Old World monkeys and the apes). Sometime in the early Oligocene, the catarrhines then split to give rise to the anthropoid apes on the one hand and the Old World monkeys on the other. The fossil record can be made to fit this general picture, and we could say that it supports this general picture. But all we really can do is describe a few of the stars—the odd finds that have come to light over the past hundred or so years in both Africa and Asia—and plug them into what we presume must have been the reality.

One of the earliest of these stars and among the best known is *Aegyptopithecus,* which has been found at Fayum and elsewhere. It was the size of a midsize monkey (about the size of a modern howler), and most people would simply conclude that it was a monkey. But it certainly was not a modern monkey. Usually it is classed simply as a primitive catarrhine.

Some creature very like it could well have been ancestral to apes and monkeys (and probably to New World as well as Old World monkeys). *Aegyptopithecus* could also be the direct ancestor of *Proconsul,* which lived around 20 million years ago, in the early Miocene. *Proconsul* too was African; the first fossil (a piece of jaw) was found in Kenya in 1909. *Proconsul* was also pretty big—almost ninety pounds (40 kilograms). Again, it had features in common with both monkeys and apes. Some believe it was already an ape and suggest that it was the ancestor (or very like the ancestor) of all the modern great apes, and of human beings. That is, one group of *Proconsul's* descendants stayed in Africa and evolved to become the gorillas, chimps, and humans; and another group found its way to Asia and evolved into orangutans. Well, it's possible. But again, the chances that this creature that we happened to find should actually be *the* common ancestor of all the big apes (and ourselves) are pretty slim.

The New World monkeys continue to mystify. It seems more or less certain that their ancestors lived in Africa, and that probably they diverged from the catarrhines way back in the Eocene. But the first sign of them, anywhere in the world, is in South America, dating from about 20 million years ago. How they got there is anybody's guess. Their ancestors might have found their way to North America during the Eocene, first crossing into Eurasia and then walking and swinging through the forested land bridge that joined Eurasia to North America. But there are no signs of early monkeys in North America, and in those Eocene days, South America was still an island, far away from North America. The most-favored idea is that monkeys first got to South America by rafting directly from Africa on floating vegetation. It's possible; this is the only way that lemurs could have gotten to Madagascar. But this explanation isn't entirely convincing. We could explain anything with these hypothetical rafts. They have the feel of the notorious literary escapes in boys' adventure stories.

The picture of primate evolution is too rough for comfort, and the piecing together is hard. It's a wonder that we know anything at all, considering how unlikely it is that fossils should form and survive, and given the huge effort required to find them. Ida, as we will see, enriches the picture to no end.

But before we move on to discuss Ida in the detail she deserves, one final footnote is called for.

The Ticklish Issue of Missing Links

It was Charles Darwin who, in 1859, first proposed publicly, formally, and persuasively that living creatures evolved from ancient, primitive ancestors that in the beginning would have been very different from the way they are now. This evolution has been continuous, with no interruptions. The change has been gradual, generally too small to notice from one generation to the next. But over time these small changes could, citing Darwin's example, turn a bear into a whale (though it now seems that whales evolved from hoofed animals). The course of evolution is shaped in large measure by natural selection, which ensures that each kind of creature, whether animal or plant or mushroom, is well adapted to its circumstances—well enough adapted to survive.

To people raised in modern biology or indeed in the post-Darwinian world, all this seems obvious. But it was not so to the intellectuals of the mid-nineteenth century. The most significant objectors were not the clergy and the theologians, many of whom took Darwin's evolutionary ideas in their stride, as many still do. Most damning was the criticism that came from fellow scientists. Some exposed technical flaws in Darwin's thinking—which in general he acknowledged and was at pains to put right. But some combined their science with theology and criticized him on grounds that were both scientific and theological.

Prominent among these scientist-theologians was Darwin's lifelong rival Richard Owen, acknowledged as the greatest anatomist in England, and probably in the world. Many of his insights from the early nineteenth century still form the orthodoxy. One of his minor but memorable achievements was coining the term *dinosaur*. He was also, by nature, authoritarian. His main criticisms of Darwin were surely wrong. But because he was obviously a great expert, and a very forceful character, he had tremendous influence.

Owen objected—on theological grounds—to Darwin's idea that life on Earth began very modestly and then gradually, and uninterruptedly, became more elaborate. This is not what Genesis seems to say. Genesis implies that all creatures were created, once and for all, in their finished form. Then there is the story of the Flood, which says that God grew tired of the creatures that He had created first and drowned them all, except for a few that found refuge on the ark. Probably

Owen did not suppose that the ark was literal. But he did think, as indeed did many late eighteenth-century and early nineteenth-century geologists, that the world in the past had suffered a series of catastrophes that wiped out the existing creatures; and then God had simply created a new batch.

If Owen had argued merely in this way, citing ancient texts, his influence surely would have been more limited. But he did not. He appealed to the fossil record, of which his knowledge was unrivaled. The fossils, he said, emphatically do not tell us what Darwin suggests they should. They do not show continuous gradual change over time. Instead, they provide a picture that seems far more in line with the account in Genesis: a succession of floras and faunas, appearing as if ready-formed and in great variety, and then wiped out by some catastrophe such as the Flood. Then another suite of creatures comes on board, apparently ready-made.

Furthermore, said Owen, if one kind of creature (like a bear) could turn by a succession of gradual changes into something quite different (like a whale), then the fossil record should contain the intermediate types. But it does not. The intermediates are missing. There are, in short, missing links. Owen said that the fossil record left Darwin's idea in tatters.

We may choose to believe for all kinds of reasons that Darwin was right—if not always in detail, then certainly in principle. Living creatures have evolved over time usually from simpler beginnings. There have been no interruptions. In general the change has been slow. But Owen's criticisms were and are valid too. If all we had to go on was the fossil record, we might be just as inclined to believe a literal interpretation of Genesis. So how do we cope with this discrepancy? And what has all this got to do with Ida?

One obvious point is that the fossil record is still very inadequate. Only a minute fraction of extinct creatures become fossilized; many of those that do are then destroyed; and although we cannot know what creatures are missing, we do know that entire sequences of rocks from entire continents are just not there—including most of the Eocene from Africa, which probably played such a part in primate evolution. We can see that the present account of primate evolution is full of gaps. We have no halfway creature between the hypothetical ancestor that resembled a tree shrew and the first archaic primates, for instance. We have no fossil that links the archaics convincingly with the first prosimians.

Yet the present record is far more complete than it was in Owen's day. Some of the missing links that Owen drew attention to have now come to light. In Darwin's day, it was indeed very hard to see how birds could have evolved from reptiles, as he proposed, since there was nothing in the fossil record that was half bird, half reptile. Then, just two years after Darwin published his *On the Origin of Species*, *Archaeopteryx* came to light at Solnhofen, Germany's other great site for fossils (along with Messel). Owen said that *Archaeopteryx* was just another bird, but it soon became clear that it wasn't. It was indeed half bird, half reptile, and one of the most spectacular missing links the world has seen.

But it is also clear that Owen's "catastrophism" was not entirely wrong. The world has suffered a series of mass wipe-outs. The Flood recorded in Genesis may well have been real; the rising waters from the end of the last Ice Age may have lingered in folk memory. The asteroid that is now thought to have wiped out the dinosaurs was even more spectacular. But few biologists today suppose that God simply started again after each mass wipeout with a whole new suite of creatures. It is far more likely that a few creatures from the old guard survived any given disaster and then flourished in the world that came after, with most of their rivals gone. This is how modern paleontologists explain the "switch" from dinosaurs to mammals at the K-T boundary—and now there is quite a lot of evidence, including fossils, to back this up.

In truth, Darwin's picture of gradual, continuous change does need modification. The world itself has changed radically and suddenly quite a few times and in several different ways since life first began. Consequently, everything has had to adjust accordingly, and the Darwinian picture of gradual, inexorable adaptation is not quite how things happened. Furthermore, although most evolutionary change does seem to take place slowly (and it appears that some creatures stay the same for many millions of years), it can be very rapid in the right circumstances. Populations of creatures can sometimes change spectacularly in just a few thousand years, and certainly in a few million. All of human history from our apish days took place in just 5 million years. But it is very easy to lose 5 million years' worth of fossils.

Finally, there is a matter of statistics. Fossilization is rare, and the animals that are most likely to be fossilized are the ones that are common and widespread. But when new kinds of creatures arise—the first birds, the first monkeys, the first human beings—obviously there are only a few of them and they are in one particular place, which may or may not be a place where fossils are likely to form. And since new life forms that find themselves in new ecological niches tend to evolve rapidly, as the first human beings did, so the first representatives of their kind quickly turn into something else.

The less time that any one kind of creature spends on Earth, the less likely it is that any individuals will fossilize.

Absolutely everything conspires against the possibility that any putative missing link will get turned into a fossil, and that that fossil will then survive and finally be discovered. All fossils are a miracle. Missing links are miracles writ large. *Archaeopteryx* was and is extraordinary. And so is Ida.

CHAPTER SEVEN

FROM THE EOCENE TO US

The paths of evolution do not run smoothly. They deviate as each generation adapts to the circumstances of its times—and also for a number of other reasons that collectively can be called chance. The fate of any particular lineage of animals—whether it changes into something else or simply stays in a rut—might depend, for example, on whether some of its more obtrusive neighbors get wiped out by an asteroid and leave it with ecological space it didn't have before.

Evolutionary paths are branched too—again, for all kinds of reasons. Sometimes a rising sea cuts a population in half, and each half evolves in its own way. Sometimes a population gets washed out to sea and winds up on some distant shore, finds itself an empty niche or a new set of neighbors, and then starts all over again, beginning with the gene pool they just happen to be sharing at the time. Sometimes different members of the same population just start breeding at different times of the year, so they never mix their genes—and so on and so forth.

As creatures change generation by generation, we see transitions from one grade to the next. Some biologists object to the idea of grade, and therefore to the idea of transition, both of which can be taken to imply progress in evolution. Evolution does not produce progress, the objectors say. But in practice, the later animals on the whole are more complex than the earlier ones: they exploit more ecological niches; have more complex societies; individually are better able to learn; and behave more flexibly. This progression isn't invariable—some lineages do get simpler over time or less skilled—but the general flow is from less complex to more complex, and from less adept to more adept. Such changes can certainly be seen to be progress of a kind; indeed, they represent the same kind of technical advances that we see in cars or aircraft or computers. On the whole, more recent computers can do more than the earlier ones in a greater variety of ways and commonly with less effort; and the same on the whole is true of more recent animals compared with their ancient ancestors.

Ninety million years have passed since our first vaguely recognizable ancestors—the first archaic primates—came into being. It's about 65 million years since the first transition from archaic to bona fide prosimians (which, among other things, were the first ones that had a postorbital bar and the goggle-eyed look that goes with it). Ida, dating from about 47 million years ago, seems to represent the first stirrings of the transition that came after that, from promisian to anthropoid.

When Ida came on the scene, the world seemed designed for primates. The group as a whole got off to a rapid start at the beginning of the Cenozoic. There were convincing-looking primates around long before there were bona fide cats or bears or hyenas, and the cloven-hooved animals had hardly put in an appearance, and the elephants and whales still looked very strange by modern standards. Yet, if you saw Ida in a modern zoo, ancient as she was, you would not think her outlandish.

The Eocene was wet and warm, and the world was practically filled with tropical, or paratropical, forest; and tropical forest is the primates' métier. Furthermore, as we have seen, the great landmass of Eurasia was joined at both ends to the great landmass of North America (although for a time Europe was largely cut off from Asia by a long intrusion of water up the middle), and then Eurasia came into contact with the great landmass of Africa. So the primates were able to migrate throughout most of the world without leaving tropical forest at all. Some of the humblest creatures were just as likely to live in China as in California. Only South America, Australia, India, and Antarctica were seriously cut off from the main primate action.

But eventually, about 34 million years ago, the Eocene came to an end. This was a huge amount of time after Ida, although not so huge when measured against the world's total history. Then the tropical forest began to retreat to its present status—basically a band of trees around the equator. Soon too, the great northern continents parted company with each other, and although some creatures (including some primates) still managed to migrate between them, there was no longer the almost total freedom to roam that was offered by the Eocene. So the corridor of tropical forest around the equator became discontinuous and focused on tropical America (the Neotropics); Central Africa; India (to some extent); and Southeast Asia. Tropical forest eventually came to Northern

Australia too, although not until the continent had drifted far enough north to come into the range of the equator (and of course it never had primates until people arrived).

Outside the forest a new kind of landscape arose, one dominated by grass. It wasn't the first time the world has seen open country, but this was the world's first extensive grassland. Most of the primates retreated to the equator as the forest retreated. Only a few came to terms with the grassland, and only a very few succeeded outside the tropics. The species that extended its horizons most spectacularly was, of course, our own. Our own rise began with grassland.

But we should ask first of all why the Eocene came to an end. Why didn't the good times — from the primates' point of view — continue to roll?

Why the World Grew Cooler

Although at first sight it might seem incredible, the Azolla Event alone was probably enough to account for the rapid cooling that brought the Eocene to an end. There is enough carbon locked in the pickled tissues of *Azolla* at the bottom of the Arctic Ocean to have reduced the CO_2 content of the whole atmosphere—enough to put a stop to the "greenhouse world" that had prevailed through the millions of years of the Eocene.

The second leading idea has to do with the chemistry of the air and rocks, and the movement of the landmass of India, which finally made contact with the south coast of Eurasia about 40 million years ago and carried on moving.

But there is one further important complication that was explained in the early twentieth century by the Yugoslav (now Croatian) mathematician Milutin Milanković, a bold thinker and one of the few to offer unwavering support to the equally bold Alfred Wegener. It had long been known (indeed, ever since Johannes Kepler in the early seventeenth century) that the Earth's orbit around the sun is not circular but elliptical. It was also known that the shape of the orbit changes over periods of about ninety-six thousand years: sometimes the orbit is almost circular, and sometimes it is more elongated. In addition, the Earth is tilted relative to the sun, and the angle of tilt varies periodically. Finally, as the Earth spins, it wobbles, like a spinning top, which is known as *precession*. These three kinds of change affect the climate, said Milanković, because they affect the Earth's distance from the sun and the angle at which the sun's rays strike the Earth. Add the three effects together, he said, and you are likely to find that the Earth will get warmer and then colder at intervals of roughly one hundred thousand years. This effect is superimposed on the general temperature of the Earth, which, as we have seen, is determined largely by the amount of CO_2 and other greenhouse gases in the atmosphere; by the layout of the continents and hence the flow of ocean currents; and by the amount of ice and hence the albedo.

In periods that are generally warm—like the Eocene—the Milanković cycles of relative warmth and relative coolness don't affect things very much. But by the time of the Pleistocene, starting about 2 million years ago, the Earth had been cooling steadily for many millions of years for the reasons we have seen and largely through the rise of the Tibetan plateau. It was so cold in the Pleistocene that the cool phase of the Milanković cycles would have been enough to trigger an Ice Age. Indeed, as the prediction has it, Ice Ages should occur at roughly one-hundred-thousand-year intervals. This means that since the start of the Pleistocene, the world should have experienced about twenty Ice Ages. The geological record says that this is precisely what has happened. The latest Ice Age ended about ten thousand years ago. At present, the world is between Ice Ages, and we will have to wait and see how things pan out over the next millions of years, as the continents continue to shuffle around and the ocean currents come and go. These things are so complicated that in detail, over time, they are impossible to predict.

But to return to our main theme. Add the rise of Tibet to the death of the Arctic ferns and we have all the mechanisms we need in order to explain why the tropical days of the Eocene came to an end, and why the world has been cooling ever since.

There has always been tropical rain forest—at least enough of it to allow primates to flourish in great variety. But at the same time, over the past 40 million years or so, we have seen the spread of more open woods and prairie and steppe and pampas as much of the world has become too cool and dry for uninterrupted forest. This new kind of landscape enabled a different suite of creatures to evolve. A few primates were among the creatures that preferred the open ground, and among them were our own ancestors.

We will come to that. First we should look at grass itself, the key player in the spread of the open landscapes—and hence in our own history.

The Rise and Rise of Grass

Grass is a flowering plant, and although flowering plants first appeared in the Jurassic, they did not properly reach their

stride until the Cretaceous. Grass looks very simple in structure, but as in racing cars and haute couture, it's the kind of
simplicity that deceives. In truth, grass is very sophisticated,
and so it was that the grass family came fairly late onto the
evolutionary scene. The very first traces of grass are from the
Paleocene, although the group probably arose first in the late
Cretaceous. Grass is not at its best in wet, hot conditions;
there it tends to be ousted by trees. But after the world became
cooler and drier, grass truly came into its own, and now it
covers about 20 percent of the Earth's surface. Obviously, it is
of huge economic importance today, for without it we would
have little or no beef or lamb. But grass has always been
important to us. The evolutionary rise of grass matters to us
absolutely, but grass has become such a powerful player in the
world's ecology primarily because, almost uniquely among
plants, it likes to be eaten—not too much but quite a lot.

Together with palms and lilies and onions and the like,
grass belongs to a group of flowering plants known as *monocots*—as opposed to cabbages, roses, peas, and oak trees,
which with many thousands of other plants are classed as
dicots. In dicots, the youngest growth is at the tips of the
plant. Giraffes nibble at the highest reaches of the acacia
trees, and cooks recommend the tips of the mint and basil. If
you nip off the tops, you destroy the bit that is growing and is
most alive, and the rest of the plant must then respond as best
it can—commonly by sending out a new shoot from a bud
farther down the stem. Either it does that or the plant dies.

But monocots grow from the bottom up. You can see this
most easily in a leek (a monocot stem, or really a bundle of leaf
bases, albeit grown largely underground). The bit just behind
the root is the tenderest, because it is the youngest—the bit

that is actually growing. As you get toward the green leaves, the tissue becomes tougher and tougher because it is older the higher up it is.

So it is with grass. The sweet young flesh is toward the bottom, as every schoolchild knew in the days before pesticides, when it was safe to pick young stems and suck on them. The top is older and tough. Left to itself, grass becomes rank at the top and simply dies, and grassland that is abandoned is always liable to be overtaken by scrub, unless it is too dry or cold for woody plants to grow. But if horses or antelopes or rabbits or some other obliging herbivore nibbles off the tops—effectively mowing it before it becomes rank—then the growing tissue beneath is liberated and can go on growing. As it nibbles, the animal destroys many of the dicots that grow among the grass—or at least ensures that those that are left are of the kind that grow very close to the ground or are otherwise specially adapted. Specialist grass eaters are known as grazers; and as the grazers nibble, the grassland flourishes at the expense of other plants. Moreover, the seed of the grass (if it grows long enough to set seed) largely passes straight through the animal, and so the grassland is spread.

On the other hand, grass doesn't like to be eaten too much. Overgrazing is always a hazard (as is evident on many an overworked farm and on ill-managed common land the world over). So the grass makes it hard for the grazers. Typically, the leaves of grasses are laced with spicules of silica, and animals that aspire to subsist on grass need strong, long teeth or they would wear away. Browsers—the herbivores that feed on the shoots of leaves and bushes—commonly have flatter teeth. Grass, furthermore, is not tremendously rich in nutrients. For reasons of physics that need not delay us here, big animals cope with poor-quality feed better than small animals do, so grazers also tend to be large; and because grass grows on the open plain, where distances are great, they need long legs so they can migrate from one place to another.

Thus, as grasslands spread after the Eocene, a whole new suite of grazers evolved in unison—cattle, antelope, deer, and some large rodents. So too, over the past 40 million years, we see the transformation of horses from the quaint little four-toed creatures that shared the forest with Ida to the mighty snorting steeds of today (although a few became smaller again and went back to the forest). Grasses got their start in the Oligocene, but the first great age of grassland was the epoch that came after that, the Miocene. Most of the great beasts and herds of the Miocene are long gone, but we can still see their relicts on the Serengeti in Africa. The prairie of North America captured some of the past glories until well into the nineteenth century, with its millions of bison and pronghorn.

The events of the deep past matter to us. The amounts of CO_2 now being released into the atmosphere are comparable with the quantity that found its way into the atmosphere at the end of the Paleocene, which finally triggered the methane rush that led to the Eocene. The only real difference between then and now is that in the Paleocene, CO_2 levels rose for reasons of geology—probably from volcanoes—and now CO_2 is rising mainly because of human activity, partly the burning of fossil fuels and partly the destruction of tropical forest. At this moment, millions and millions of tons of methane are trapped in clathrates beneath the ocean floor. Soon, many scientists predict, the clathrates will melt. Then the methane will be released. Then, well, we will just have to wait and see. In general terms, we could be back to an Eocene world, though with a very different set of creatures on board—including us.

The cooling that followed the Eocene produced the kind of conditions that, many years later, enabled the descendants of Ida (or some creature very like her) to abandon the forest and take to a life in open country.

Ida seems to represent the first stirrings of the anthropoids. By the end of the Eocene, the anthropoids were firmly on board. They were diverse, and they were already quite large. By about 30 million years ago—the Early Oligocene—the anthropoids had largely replaced the prosimians as the leading primates over most of the world. Clearly, the bulk of their history has taken place in the post-Eocene world, a cooling world of retreating forest and advancing grassland. But what do we actually know about the post-Eocene primates?

From Ida to the Hominoids

The short answer to this question, What do we know? is, as ever, Not much. There are a huge number of fossils—many hundreds—from Africa, Europe, and Asia; there are none in South America from before 25 million years ago; and there are none at all in North America from before very recent times, when South America finally joined up with North America via the Isthmus of Panama. Many animals late in the Tertiary migrated to and fro between Alaska and Siberia via the Beringian land bridge, which opened from time to time. But in post-Eocene times, this land bridge formed only in periods of extreme cold, when the sea level went down, and since primates on the whole cannot make a living in tundra, this land bridge was of no use to them. So all the many hundreds of fossils that help us track the rise of the modern anthropoid groups come from Africa, Europe, and Asia. They are just a series of snapshots, and different biologists arrange these snapshots in different ways and present various stories, each of which they tend to defend with vigor.

So let us look first at the snapshots: the factual evidence.

What Do the Fossils Tell Us?

None of the post-Eocene snapshots would qualify as holiday snaps or as newspaper cuttings; they are more like fragments of old postcards found in the basements of bombed-out buildings. We can build wondrous tales from the backs of them, reconstruct life histories of people whom we have no chance of meeting and who have left no known relatives, and up to a point those tales will be valid. A postcard sent from 1930s Italy tells us that the recipient knew somebody rich enough to take a holiday in Italy between the world wars—meaning they were probably at least middle class and comfortably off and so forth. But in the end, it's all speculation.

Many paleontologists feel that most of the really important steps in primate and human evolution must have taken place in Africa. They give several reasons for this, some of them more convincing than others. One is that Africa now has the greatest variety of primates in terms of number of species and range of body types: lots of monkeys; three great apes; and prosimians in the form of bush babies and pottos. The lemurs are confined to Madagascar, but it seems that they too originated in Africa. Of course, some or all of those creatures might have originated somewhere else, wandered into Africa (over vast periods of time), and then flourished while dying out elsewhere. But the evidence now before our eyes suggests that Africa at the very least was a highly significant cooking pot.

A great many fossils support this claim. Egypt's Fayum Depression, about fifty-six miles, or 90 kilometers, southwest of Cairo, has yielded a wondrous collection of anthropoids from the Eocene-Oligocene boundary, around 34 million years ago, showing that they were already a varied lot at that point. Charles Darwin also suggested that Africa was probably the site of human origins. After all, he said, our two nearest living relatives, the gorillas and chimpanzees, are still there—and it's at least a reasonable bet that we all three share a common ancestor who also lived in Africa. That common ancestor has not yet been found (or at least none that really seems to fit the bill), but the earliest-known members of the human family—hominids—have been found there, so an African origin for humans now seems as near to certainty as this subject can provide.

But we have to be careful. Seek and—sometimes—ye shall find, and it's always possible that the evidence so often favors Africa because that is where most of the digging has been done. Certainly for the early part of anthropoid evolution, Asia has its fans too. We must keep an open mind. Europe, the home of Ida, is peripheral to both great continents. But it also provides a significant corridor between the two, and surely it was a significant launchpad for at least some groups.

In practice, the snapshots of ancient anthropoids come from both Eurasia and Africa. Either could be the scene of most of the action, or both could be equally important. Certainly, even if those ancient types were basically stay-at-homes, they had plenty of time to move from one continent to the other and back again—and then back again. After all, the period relevant to this part of the story extends from the Late Eocene, just over 34 million years ago (when it's clear that the anthropoids had already come on the scene), to the boundary of the Miocene and Pliocene, around 5 million years ago, when the first bona fide hominids were around. (That's fifteen hundred times longer than the time that has passed since the birth of Christ.)

The term *hominid* means "member of the family Hominidae," which of course includes our own genus, *Homo*. Unfortunately, different biologists define the Hominidae in different ways, which means that the adjective *hominid* also has different meanings. Traditionally, the Hominidae family was taken to include *Homo* and its immediate relatives, such as *Australopithecus,* whom we will meet shortly. Chimpanzees were placed in a different family. But in recent years it's become clear that there is very little genetic difference between humans and chimps and very little case, therefore, for placing them in separate families. So, many biologists these days include chimps in the Hominidae, which should mean that they are hominids too. In this account, for reasons of clarity, I am using the word *hominid* in the traditional sense—to mean *Homo* and its immediate relatives as opposed to the chimps and the other apes. But throughout the book, as we have seen, the term *anthropoid* is used to include all the apes and all the monkeys, both Old World and New World; in other words, anthropoid as opposed to prosimian.

These ancient anthropoids would have been inhibited in their travels only by lack of suitable terrain. On the whole, the ancient types were still forest animals, or at best were adapted to open woodland, and they all probably needed a hot climate. They could not travel between Asia and Africa unless they had suitable habitat all the way. Hiring a camel to cross the occasional desert or a boat to cross some transient lake was not an option. But they would not have needed a continuous corridor of trees at any one time. We can envisage that as the landscape changed—both because of climate and through continental drift—they could from time to time have extended their range in one direction or the other as woodland rose up where there was none before; and then at some later stage, the terrain in front of them would also start to acquire trees, and they could move on. Thirty million years is ample time for major forests to come and go several thousand times over. After all, the great boreal forests of present-day Siberia and Canada and the tropical forest of Queensland, which seem to have been there forever, date only from the last Ice Age, about ten thousand years ago. You could fit the entire lifetime of those forests into the vast span of time between the end of the Eocene and the start of the Pliocene—three thousand times over.

In short, although in general it is good in science to keep all explanations as simple as possible, we should not slavishly assume that everything interesting must have taken place in Africa, or discount theories that involve a two-way flow between Africa and Asia, just because they are more elaborate. There has been plenty of time and plenty of opportunity for very extraordinary things to happen.

We already know enough to see that Asia could be very important. *Eosimias* from China is thought by some to be the clue to all anthropoid evolution. Perhaps even more striking (though still equally controversial) are two creatures from

177

Myanmar (formerly Burma): *Amphipithecus* and *Pondaungia*.
They are of impressive size—about as big as a gibbon—and
some primatologists claim that they are very definitely anthropoids. Yet they date from around 40 to 44 million years ago,
deep in the Eocene, which for anthropoids would make them
very ancient indeed. There is also *Siamopithecus* from Thailand (or Siam) from about the same period, which some say
is related to *Amphipithecus* and *Pondaungia*. Other paleontologists say that these creatures are not anthropoids—only
fragments of jaws have been discovered so far—and some
have said they are not even primates. But if they do prove to
be anthropoids, then scenarios of anthropoid origins based
in Africa would be seriously challenged.

But Africa has yielded a host of wondrous anthropoid
fossils from the Fayum, thanks mainly to the extraordinary
work of Elwyn Simons of Duke University from 1961 until
the end of the century. You would not think, looking at the
Fayum today, that it's a good place for primates, any more
than you would suppose that Messel was a good place for
primates. It is the most barren of deserts, on the edge of
the Sahara—not the romantic kind with rolling dunes and
moody Italian actors pretending to be Arab, but stony, the
kind that the Bible is apt to call wilderness. Many have felt, in
a cautious vein, that it could never have been wetter than, say,
the modern-day Sahel, the stretch of land to the south of the
Sahara, which has just about enough rain to grow some crops
but is basically scrubland. But Dr. Simons has helped show
that it was once tropical forest, and clearly wetter. He developed an extraordinary technique for revealing the Fayum
fossils: remove the surface stones to expose the sand beneath;
then go away and come back next year. In the meantime, the

desert wind does its work, blowing away the fine sand and exposing an astonishing range of fossils beneath that can just be picked from the surface. They show beyond doubt what riches there once were. Most revealing of all, perhaps, and exposed with miraculous delicacy by the Sarahan wind, are the needle-thin foot bones of jacanas, or lily-trotters, so called because with their long, spidery feet they can run over the lily pads as if they were on a playing field. Where there are jacanas, there are tropical forest lakes. There are the fossils of trees and lianas too, just to emphasize the point. It was very like ancient Messel, in short, though even hotter.

The Fayum fossil primates are bona fide anthropoids, but the earliest ones, at least, predate the split of the New World monkeys from the Old World catarrhines—and also, of course, predate the split of the catarrhines into the apes on the one hand and the Old World monkeys on the other. Truly, then, they offer a privileged insight into the origins of modern primates, as if we could rediscover the hunting-gathering people from whom we are all descended. The range of types is impressive. About a dozen different anthropoid genera have been unearthed, some of which include several species. Some of them were found long before Elwyn Simons turned up in 1961 (which was why he went there in the first place), but he has added significantly to the list.

Most of the Fayum primates are split between two families. The older family, and the more primitive, is the Parapithecidae, which retains the primitive number of premolars—three on each side, on the upper and lower jaws, as are still seen in the New World monkeys. The more recent and more derived family is the Propliopithecidae, whose members have only two premolars per side per jaw. Some taxonomists also recognize a third family, the Oligopithecidae, although this one is clearly very close to the Propliopithecidae. It would take too long to describe them all (and in any case, many are known only from fragments of jaw, so there isn't much to describe), but it's worth looking at a few.

Apidium was the first-known member of the older family of Fayum primates, the Parapithecidae—and indeed, the first Fayum primate of any kind to be described. That was back in 1908, by Henry Fairfield Osborn, whom we will meet later. He called it *Apidium phiomense*. In 1961, his first year in the Fayum, Elwyn Simons discovered an earlier and so far much rarer species of the same genus, which he called *Apidium moustafai*, after one of his coworkers. Now a third species has been added. Fossils from the three *Apidium* are known from between 36 million years ago—Late Eocene—and 22 million years ago—late in the Oligocene. The later ones, therefore, lived on well after the split between the New World platyrhine monkeys and the catarrhines. They looked like small monkeys, running on all fours along the branches and leaping between the trees. Enough fossils are known to show that they were sexually dimorphic, the males much larger than the females and with big canines. These are the hall-marks of a polygynous creature—a male battling to achieve sole command (or at least shared command) of a harem.

The creature for whom the more recent family, the Propliopithecidae, is named is *Propliopithecus*. This too was first described early in the last century, by a German paleontologist in Stuttgart, Max Schlosser. He acquired the fossil from the same collector (also a German) who was also then providing fossils for Osborn. *Propliopithecus* wasn't large—a mere fifteen or sixteen inches (40 centimeters)—but it is said to have had the same kind of shape as a modern gibbon. Schlosser called it *Propliopithecus* because it resembled the gibbonlike and much larger *Pliopithecus,* which was already well known from France and Switzerland but lived much later in the Miocene and Pliocene.

Very similar to *Propliopithecus,* and placed in the same family, is Elwyn Simons's most famous find, *Aegyptopithecus zeuxis.* It lived at the time of the Eocene-Oligocene boundary, 33 to 35 million years ago. Although *Aegyptopithecus* is sometimes called the "Dawn Ape," it actually resembled a modern howler monkey in size and general shape, though it had only two premolars on upper and lower jaws, like a catarrhine. It was probably an arboreal quadruped (walking along the branches on all fours), and its large opposable big toe gave it a powerful grip with its hind feet, probably so it could browse on fruit and leaves. Again, it was sexually dimorphic, a creature of harems. But it had a surprisingly small brain.

The third family, Oligopithecidae, took its name from another of Elwyn Simons's finds, *Oligopithecus savagei.* It dates from the early Oligocene, about 32 million years ago, and was midsize by monkey standards: about three pounds (1.3 kilograms). Since it is known only from a single jawbone, there isn't much to add.

Catopithecus, from the same family, is far older, dating from 37 million years ago, well back in the Eocene. *Catopithecus* was the size of a modern marmoset, and as one of the oldest of all the Fayum anthropoids, it's therefore one of the oldest anthropoids that is universally recognized as an anthropoid.

So fossils are indeed known from Asia, notably from

Myanmar and Thailand, and some claim that they are bona fide anthropoids from the late Middle Eocene, more than 40 million years ago. There is a whole range of anthropoids that nobody doubts are from North Africa, namely Egypt, but they date from much later—most around the Eocene-Oligocene boundary. Time and more fossils will perhaps tell us what really went on; whether the Fayum fossils truly represent the start of the anthropoid line or are, as some in effect are claiming, just a sideshow, with the real action taking place in Asia much earlier. The latter is a minority view, but we will just have to wait and see. Meanwhile, we can at least be certain that anthropoids were well into their various strides by the early Oligocene, and we should ask what happened after that.

From the First Anthropoids to Near Modernity

Science is always a balance between theory (how we think things work, or what we think things must be like) and empirical evidence (what we take to be the facts of the case). Ideally, the theory and the facts match up perfectly, but usually the theory and the facts (or different sets of facts) are in a state of tension, which is why research can never stop, and why science never stops being interesting.

We can guess in broad terms what must have happened in primate evolution with very little reference to the fossils at all. For example, if we compare the DNA and other molecules from living types, we can judge, roughly, how long ago it is since they shared common ancestors. Then we can compare what the molecular evidence tells us with what the fossils tell us. For instance, comparison of DNA tells us that the New World platyrhine monkeys parted company with the Old World catarrhines about 40 million years ago. But the fossil anthropoids from the Fayum typically combine features both of platyrhines and catarrhines, and they generally date from around 34 million years ago. Hmm. If we take the Fayum fossils at their face value, we must conclude that catarrhines and platyrhines parted company around 30 million years ago, in the early Oligocene. But the molecules—or some of them, at least—suggest an older date: the Middle to Late Eocene. Most agree that the ancestors of the New World monkeys must have arisen in Africa and ferried their way across the Atlantic to South America on floating vegetation. It is not a totally convincing explanation, but it is possible, and without something better, it will have to do.

Similarly, it seems that the anthropoid apes (which are all Old World) broke away from the Old World monkeys somewhere between 33 million years ago—the date of *Aegyptopithecus*—and 20 million years ago—the date of *Proconsul,* who was a bona fide ape (of whom there will be more later). The "lesser apes" of Asia, alias the gibbons, are presumed to have parted from the "great apes" around 18 million years ago. The orangutans, now exclusive to Asia, parted from the African great apes probably around 13 million years ago. In Africa, the gorillas apparently broke with the chimpanzees-plus-hominids around 7 million years ago; and very soon after that, the chimps and the humans broke apart and started to evolve in very different ways.

The task now is to fit real fossils into that grand scenario. To some extent, this has been done—or at least the work is well in progress. A great many genera and species of many different primate families are now known from the Oligo-cene, Miocene, and Pliocene—from Africa, Europe, and Asia, and indeed, from the end of the Oligocene, from South America. But again, most of the fossils that are known are only fragments, and they inevitably come from those few sites that happen to contain good fossils, and they do not tell a coherent story. Or at least they can be made to tell a great many different stories of equal coherence, leaving us with no obvious way to choose from among them.

We should look at a few of those Oligocene-to-Pliocene anthropoids just to get a flavor of what was out there.

Roughly at the point of division between anthropoid apes and Old World monkeys is *Proconsul,* dating mainly from the Miocene in Africa—27 to 17 million years ago. It has fea-tures of both: a deep narrow chest like a monkey but, among other things, a more ape-size brain and no tail. It walked on all fours, but on the palms of its hands like a monkey, not on its knuckles like an ape. Some feel it could be the first of the anthropoid apes—sufficiently primitive but sufficiently ape-like to have given rise to all of the modern great apes: chimps, gorillas, and orangutans. Others feel it is probably just a rela-tive of the first-ever ape: a cousin, and an evolutionary dead end. There is no way of knowing, but the putative common ancestor of apes and Old World monkeys presumably was something like *Proconsul.*

The first *Proconsul* was discovered in Kenya in 1909, and now at least three species are known, and some say there were probably more. They were substantial beasts, the small-est species weighing about twenty-five pounds (10 kilograms) and the biggest about ninety pounds (40 kilograms), the size of a small chimpanzee. The name *Proconsul* was coined in the 1930s. It means "before Consul"—Consul being at the time a popular name for chimpanzees. The Folies Bergère in Paris had a chimp called Consul in 1903, and London Zoo had another of the same name in the 1930s.

Dryopithecus, another Oligocene-to-Pliocene anthropoid from the Middle Miocene (12 to 9 million years ago), is known from East Africa, Europe, and Asia. The first was found in France in 1856, and others have turned up in Hungary, Spain, and China. It probably evolved first in the Rift Valley of Africa and then spread outward.

Dryopithecus wasn't a huge beast, apparently around two feet (60 centimeters) long. It was primarily apelike but again had some features more reminiscent of monkeys. Like a monkey, it walked on the palms of its hands rather than on its knuckles. But it had a broad, flat chest like an ape's, and its lumbar vertebrae were stiff like an ape's (and like ours), not flexible like a monkey's. Most significant, its arms suggest that it could swing from the trees like an orangutan—the oldest ape that is known to do this. The thin enamel on its teeth and the rounded, apelike cusps on its molars suggest that it ate fruit with some leaves. All in all, it seems close to the ancestry of the orangutans.

Sivapithecus, Ramapithecus, and *Kenyapithecus* should be discussed together because in truth they might all be the same beast. Whether they are or are not all the same, they were a widespread, presumably ecologically important set of impressively large apes—from chimp to orangutan size—that lived in the Middle Miocene, probably from about 17 to about 12 million years ago. Thus, they help to fill the highly significant time gap between the obviously primitive *Proconsul* and the moderns: the orangutans on the one hand, and the chimps, gorillas, and humans on the other.

Ramapithecus was the first to come to light, in the early 1930s, in Nepal. Its finder, G. Edward Lewis, claimed that its jaw was more like a human's than that of any other fossil ape then known. In the 1960s, no less than Elwyn Simons declared that it probably was the oldest known bona fide hominid—directly on the human line. His student David Pilbeam, now a distinguished professor and curator of paleoanthropology at Harvard, supported him in this. It was the overall shape of the jaw that made it seem humanlike; it seemed parabolic in general shape, whereas the lower jaws of apes are more V-shaped. But then, at that time, no one had an intact jaw of *Ramapithecus*. They had to infer the overall shape by joining bits together.

In the 1960s too, just to stir the pot a little more, Louis Leakey, father of Richard Leakey, found in East Africa yet another human-looking ape from the Miocene, which he called *Kenyapithecus*. This was older than *Ramapithecus*—about 15 million years old—so surely this, rather than *Ramapithecus*, was the most ancient known member of the human lineage. Simons, however, opined that *Kenyapithecus* was the same as *Ramapithecus*. The fact that it lived on a different continent really isn't a problem, as we have seen.

Simons and Pilbeam's claims for *Ramapithecus* seemed highly plausible because, in the early 1960s, paleoanthropologists believed for various reasons that humans had split from the other apes at a very early date: at least 16 million years ago or even, some said, way back in the Oligocene. If that were the case, then an ape that lived 14 million years ago could already be on its way to becoming specifically human. But in the late 1960s, biochemical evidence came on board. Allan Wilson (a biochemist) and Vincent Sarich (an anthropologist) at the University of California at Berkeley began comparing blood proteins (specifically albumins) from different animals. If two animals have very similar albumins, it means they are closely related genetically, which means they must have shared a common ancestor very recently; and if the albumins are different, then their common ancestor must have lived longer ago. They found that humans and chimps have remarkably similar albumins; so similar that they must have shared a common ancestor about 6 million years ago, and certainly not more than 8 million. Scientists at first said this was nonsense, but the biochemical evidence grew stronger and stronger. If this was so, then *Ramapithecus* was clearly too old to be specifically hominid. It dated from well before the apparent split between humans and chimps. Then, in 1976 David Pilbeam found a whole *Ramapithecus* jaw. It proved to be V-shaped and was clearly more ape than human.

Sivapithecus, like *Ramapithecus,* was discovered in the 1930s in Asia, and three species are now known, from India, Pakistan, and Turkey. It lived somewhat later than *Ramapithecus,* around 12.5 to 8.5 million years ago. Overall, its body was quite chimplike; it could certainly climb, but it probably spent a great deal of time on the ground. It likely ate a lot of seeds and rough grass in open Miocene woodland and the surrounding savanna; its heavy, thick-enameled molars certainly seem suited to that. But as David Pilbeam confirmed when he found part of a jaw and face in 1982, its face was more orangutan-like. Now it's thought that the orangutan lineage diverged from the lineage of the chimps, gorillas, and humans around 16 million years ago, and that *Sivapithecus* is probably on the orangutan lineage. At the same time, it seems more and more likely that *Ramapithecus* is really the same as *Sivapithecus*. Indeed, it seems that *Sivapithecus* was sexually dimorphic—and the so-called *Ramapithecus* could be the females. Since *Kenyapithecus* is probably the same as *Ramapithecus,* this means they should all be called *Sivapithecus*. The name *Sivapithecus* takes precedence because it was the first to be described. And, to repeat, *Sivapithecus* and its associates were not Miocene humans, since they were on their way to becoming orangutans.

Sivapithecus by whatever name was clearly highly successful. The genus lived a long time and over a very wide area, and it probably gave rise to several descendant lineages. The orangutans are probably one of them; so too, probably, is the biggest primate that has ever lived: *Gigantopithecus*.

Gigantopithecus was a kind of superorangutan (though surely more at home on the ground) that lived in China, India, and Vietnam, and in surprisingly recent times: from about 1 million to around three hundred thousand years ago. Its bones first came to light in 1935. In ground-up form, the bones are still found in the shops of Chinese apothecaries, and in 1955 a consignment of "dragon teeth" from China included forty-seven that actually came from *Gigantopithecus*. Three species are known, but of course there might have been more. *Gigantopithecus* really was huge. The biggest, *Gigantopithecus blacki,* stood almost ten feet (about 3 meters) tall. It must have weighed around half a ton—two or three times heavier than a modern gorilla and about five times the size of a modern orangutan. The females were only half the size of the males, however, suggesting that *Gigan-*

topithecus, like the gorillas, was polygamous. Its teeth were big and thickly enameled and clearly adapted for a harsh diet; perhaps, like the giant panda, it specialized in bamboo, though it probably ate fruit and other plants as well. Many have suggested that the legends of the yeti are folk memories of *Gigantopithecus,* and this is eminently plausible. Somewhat less plausible, but wonderfully intriguing, is the idea that *Gigantopithecus* is still out there.

But at least one known Miocene ape seems to have been close to the ancestral line of the African apes and ourselves. This is *Nakalipithecus nakayamai,* first described by Japanese paleontologists in 2005 from Nakali in Kenya. It dates from around 8 million years ago, late in the Miocene, and close to the time when gorillas split away from chimps and humans. Perhaps too it was related to *Ouranopithecus* from Greece and Turkey—although some have suggested that *Ouranopithecus* could be descended from *Sivapithecus.* The heavy teeth of *Nakalipithecus,* with their thick enamel, suggest a diet of harsh vegetation.

Finally, we might note in passing how successful, widespread, and diverse the great apes were in the Miocene. There were at least seven genera, many of which contained several species. Nowadays the gibbons and siamangs are still diverse, but they are classed as lesser apes. The great ape group is now down to four genera, with very few species: *Pan* (the chimp and the bonobo, two species); *Gorilla* (two species, but very similar); *Pongo* (the orangutan, two species, but again very similar); and *Homo* (one species). All except humans are confined to tropical forest and endangered. In contrast, the monkeys have thrived and diversified since the Miocene. We tend to feel that the apes are superior to the monkeys. They are bigger, on the whole they seem cleverer, and besides, they include ourselves. But then, as Ecclesiastes reminds us, "The fight is not to the strong. Time and chance happeneth to all."

So how did a late Miocene ape evolve into the first hominid? The picture is becoming steadily clearer. But the search got off to a shaky start, and the main problem was one we have discussed before: that of search image.

The Meandering Path to Humanity

Anatomically, the thing that makes modern humans very different from all other animals is the size of our brains relative to our bodies. The modern human's brain averages around 1350 milliliters, compared with about 450 milliliters in chimpanzees, our nearest living relatives.

What we actually *do* with those brains that makes us different from other animals has not been so easy to define. Simply to say we are cleverer doesn't seem to get quite to the essence. Biologists used to say that human beings were the only animals that used tools—but this is clearly untrue, since chimpanzees use rocks to crack nuts, and some finches use thorns to probe grubs out of crevices, and so on. Biologists then shifted the ground and said that we are the only animals that *make* tools. But this is not true either. As Jane Goodall showed, chimpanzees modify twigs so that they can use them to lure termites out of their nests for a nutritious snack, and now we know that New Caledonian crows make hooks out of thorns to dig out food (and in the laboratory, they will make hooks out of wire or whatever is around).

René Descartes in the seventeenth century said that what makes us different is that we speak. But that doesn't seem to work, because it is now obvious that many mammals and birds have a wide vocabulary and communicate a great deal. Vervet monkeys make different noises to warn their fellow vervet monkeys about different kinds of predators, such as eagles and snakes. Chickens have about twenty different calls for different purposes.

Yet it seems that Descartes was on the right path. Speech and all that goes with it is the key. First, unlike chickens, we can invent new words at will to describe whatever comes along, and actions and states of mind as well. Second, on a purely physical point, we have a larynx low in the throat that enables us to articulate rapidly and accurately—to speak at speed. Chimps by contrast, like virtually all other mammals, have their larynxes high in the throat and in general make only strangulated sounds. Most important, however, as the American philosopher Noam Chomsky pointed out, our language is underpinned by rules of syntax. This enables us to manipulate our array of words at will to convey just about any thought of which we are capable.

This in turn means that we can share our thoughts, rapidly and accurately, and that as thinking beings, none of us are on our own. Now that writing has been invented, we can in principle share the thoughts of all other human beings that have ever lived, or at least the ones who wrote things down. Taken individually, we may not in practice be able to outsmart all other creatures—certainly not in all circumstances. If it was a straight battle for survival in the wilderness between a human being and a hyena, it is not self-evident that we should put our money on the human being—and it's not just a matter of physical strength. Hyenas are astonishingly aware and streetwise. Yet they cannot do what we can do. They cannot seek advice and learn woodcraft from the commandos. Intellectually speaking, hyenas are more or less on their own, but human beings can plug themselves into the great collective brainpower of humanity. The advantage is tremendous, and it is enough to explain why we have become the commonest large mammal on Earth, by far, while our nearest genetic relatives, the other great apes, are endangered or nearly so. Each of us on our own, without cooperative thinking and learning, is pretty feeble and seriously inept.

But how did we get to be the way we are? Who were our ancestors? This is where search image comes in. What should paleoanthropologists be looking for? Presumably the first hominids would show some features of apes, since they had only recently diverged from the chimps. Also presumably they must have some special features of humans too, or we would not suppose that they had anything particular to do with our own ancestry. But what kinds of features would they possess from each?

Because human beings are above all brainy, some of the leading post-Darwinian paleoanthropologists who first set out to answer this question assumed that our own evolution must be "brain-led." The creature they felt we should be looking for—the "missing link" between apes and humans—would surely possess an apelike body with a large-domed skull. Only after our first unequivocal ancestor had acquired its big brain would it be in a position to abandon its apish physique and the strength that went with it.

The logic was inexorable. But the search image became an idée fixe, and this led to one of the most bizarre incidents in the history of science.

In 1912, in a gravel pit at Piltdown in Surrey in southern

England, there came to light what seemed to be one of the all-time star fossils: part of a skull and lower jaw from what seemed like the ideal ancestral hominid. The cranium was huge—so it was a big-brained creature—and the jaw was distinctly apelike. It was assumed (although no other bones were found) that it had an apelike body to go with its apelike face.

The experts of the day included Sir Arthur Keith, Sir Grafton Eliot Smith, and Sir Arthur Smith Woodward, and it was exactly what they were looking for: a creature led by its human-size brain out of its apish past. Furthermore, it had been found in England, which was at the time the center of the world's greatest empire, so it was exactly where they expected to find the first human being. Darwin had suggested the first humans lived in Africa, and these gentlemen were good Darwinians, but they did not suppose him to be right in every last detail. From America came support from Henry Fairfield Osborn. They called the creature *Eoanthropus dawsoni*. *Eoanthropus*, of course, means "Dawn Man"; and *dawsoni* was in honor of Charles Dawson, a local amateur collector who first produced the find—acquired, so he said, from a workman on the Piltdown site.

Some scientists, however, were not convinced, and quite rightly, for in 1953, studies at the Natural History Museum in London showed that *Eoanthropus dawsoni* was a hoax. The skull came from a modern human. The jaw was from an orangutan, crudely filed to eliminate the bits that would have given the game away. Then it had all been dyed. No one yet knows for sure who perpetrated the hoax, but Dawson is the chief suspect. It fooled the leading experts of the day because it had been tailor-made to match their preconceptions.

The tale is very sad. Keith, Smith, and Smith Woodward

were truly accomplished, and all were made to look foolish (though only Keith survived to see the hoax exposed). They surely would not have been fooled if they had not been so sure in advance that the mocked-up *Eoanthropus* was the kind of creature they should be looking for. Sadder still, though, was the way that in the light of Piltdown, they and others rejected the announcement, in 1924, of the real thing: *Australopithecus africanus.*

Australopithecus africanus came from a lime pit in Taung, South Africa, and was given to a young (just thirty-one years old) émigré Australian anatomist named Raymond Dart, who had just started work in a hospital nearby. There was only a bit of the skull, including part of the face and cranium. The creature was clearly immature—probably equivalent to a five-year-old child—but there was enough to infer that in adulthood it would have stood about three feet (1 meter) tall; and to infer that the brain for such a creature would have been of chimplike proportions—and later adult specimens suggest around 485 milliliters. Most revealing of all, though, was the foramen—the hole that marks the entry of the spinal cord into the head. In apes such as chimpanzees, the foramen is very far back in the skull because chimps in general hold their bodies horizontal, with their heads thrust forward. In humans, the skull is perched on top of a vertical vertebral column, and the foramen is tucked well underneath. In this new skull, the foramen was underneath, so it was reasonable to infer that this small creature was upright too. Largely because of its supposed upright stance, Raymond Dart suggested it as a plausible link between apes and humans. His name for it, *Australopithecus,* means "southern ape." The particular one that he named, that is, the "type specimen," is often referred to as the "Taung Baby."

But the gentlemen back in England were not impressed. They already had Piltdown. Equally to the point, *Australopithecus* was precisely what they did not expect: it was not a creature with a human-size brain on an apish body, but one that still had a fairly ape-size brain with a presumably upright, human-style body. Furthermore, despite their Darwinian credentials, they did not expect it to have lived in South Africa. Neither, apparently, did they like the idea that it had been found by an Australian, a colonial. In the words of Phillip Tobias, professor emeritus of paleoanthropology at the University of Witwatersrand in Johannesburg, and a disciple of Raymond Dart, the Taung Baby was "the wrong fossil in the wrong place and was found by the wrong man" — at least as far as Britain's learned professors were concerned. But after the ghost of Piltdown was finally laid to rest in 1953, the genus *Australopithecus* began to take its proper place as the true ancestor of *Homo*.

Australopithecus africanus itself was not our direct ancestor: the Taung Baby lived too recently—around 3 to 2 million years ago; and by 2 million years ago, creatures that some feel qualify as the first *Homo* were already on board. Much more convincing as a possible direct ancestor was and is *Australopithecus afarensis,* which was first described by Donald Johanson and his colleagues in 1974 in the Afar region of Ethiopia. This is the famous "Lucy," so called because at the time of her discovery, the Beatles' "Lucy in the Sky with Diamonds" echoed around the Johanson camp. Until Ida, she was probably the world's most famous primate fossil, not least because there was so much of her, including more than half the skeleton. She was older than Dart's Taung Baby; she and later finds show that A. *afarensis* lived from around 3.9 to 3 million years ago. She would have been a bit taller than A. *africanus*—A. *afarensis* was three to five feet (1 to 1.5 meters) tall—but her brain was about the same size as the Taung Baby's, though her face was more apelike. Most important, though, enough of her leg bones remained to show that they would have pointed straight downward, that she did indeed walk upright, confirming what Dart had guessed from the Taung Baby's skull. It still wasn't clear whether she walked heel and toe like a modern human, or more on the sides of the feet like an ape. But in the late 1970s, Mary Leakey, wife of Louis, found the footprints of three A. *afarensis* preserved forever as they strode across volcanic ash at Laetoli, Tanzania—ash that fell around 3.6 million years ago. This showed very clearly that they walked very well: heel and toe like modern people.

More australopithecines have been found, of at least half a dozen species, various permutations of which would have lived together at any one time, side by side. Clearly, some of them were gracile and some were more robust—or at least had heavier jaws and teeth for crunching rough vegetation. The robust ones are commonly given their own genus, *Paranthropus* (and some of them in the past were given other names, such as *Zinjanthropus*). The early ancestral tree of the hominids, as with most creatures, was quite bushy.

An even older hominid genus has been found too: *Ardipithecus*. The first to be found and the earliest, *Ardipithecus ramidus*, again from Ethiopia, was at first dated to around 4.4 million years but has now been redated at around 5.8 million years—very close to the presumed split between hominids and apes. Even older, at around 6 million years, was *Orrorin,* found in 2001, from Kenya; and earlier still was *Sahelanthropus,* at 6 to 7 million years, found in Chad in 2002. So, slowly and painstakingly, though not without controversy, the picture from the late Miocene and into the Pliocene is beginning to emerge. Alas, as far as we know, there is no fossil site that can show us life in Miocene Africa in the way that Messel reveals the joys and tribulations of Eocene Germany.

From Homo Habilis to Homo Sapiens

Rolling the clock forward again, it seems that the genus *Homo* evolved from one of the gracile australopithecines something over 2 million years ago. The oldest widely (though not universally) accepted member of the genus *Homo* was *Homo habilis,* first described by Louis Leakey, John Napier, and Phillip Tobias in 1964 from Olduvai Gorge. Again, it was short by human standards—not much more than three feet (1 meter)—but its brain was much bigger than an ape's, at 600 to 750 milliliters. Its discoverers gave it its name, which means "handy man," because its bones were accompanied by stone tools—crude by later standards (later stone tools can be exquisite), but tools nonetheless. They dated this first specimen at about 1.8 million years, but other, similar types have been found since, giving a total range of from around 2.4 to 1.5 million years ago. The later types, however, are varied and commonly supposed these days to belong to more than one species. The ones that look more like australopithecines are sometimes called *Australopithecus habilis,* and some of the more human-looking ones are called *Homo rudolfensis.*

From about 1.8 million years ago we see the beginnings of a range of hominids that are generally said to be *erectus grade*. In height and build they were like modern humans, and their brains seemed to get steadily larger: the earliest ones at around 900 milliliters, the later ones at around 1100 milliliters. Somewhere along the line, erectus-grade people learned to harness fire; this was a hugely significant step in the evolution of human beings, not least (though somewhat later) in the evolution of agriculture. The hominid that is officially called *Homo erectus* was the first to leave Africa. One of the Asian types was first found in China in the 1920s and was duly called "Peking Man." The erectus-grade hominids who stayed in Africa are commonly called *Homo ergaster*. *Homo erectus* apparently lived on until about three hundred thousand years ago, which means that for a very long time it was contemporary with the first *Homo sapiens*. Again we see the bushiness of the hominid family tree. There is no God-given law that says that *Homo sapiens* was or is the only bona fide species of human being.

One final possible twist in the erectus story. In 2003, an almost complete skeleton was found on the Indonesian island of Flores dating from about eighteen thousand years ago and not yet completely fossilized. It was human, no doubt about that, and it was female, probably around thirty years old at the time of death. It was also tiny—not much more than three feet (1 meter) tall. It also had a smallish brain, even relative to its body size. Its discoverers called it a new species, *Homo floresiensis*. Others since have said that she was just some kind of microcephalic dwarf, or that she and her cohabitants of Flores were stricken by disease. But more, similar bones have been found, some dating from as recently as thirteen thousand

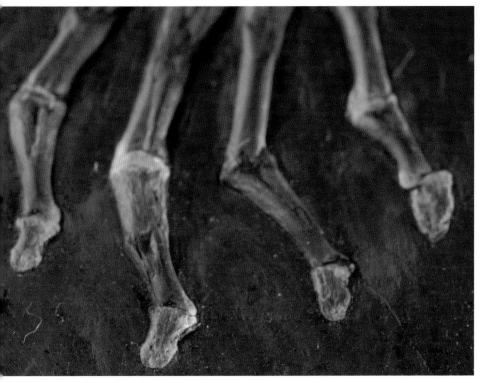

A closer look at the nails on Ida's foot.

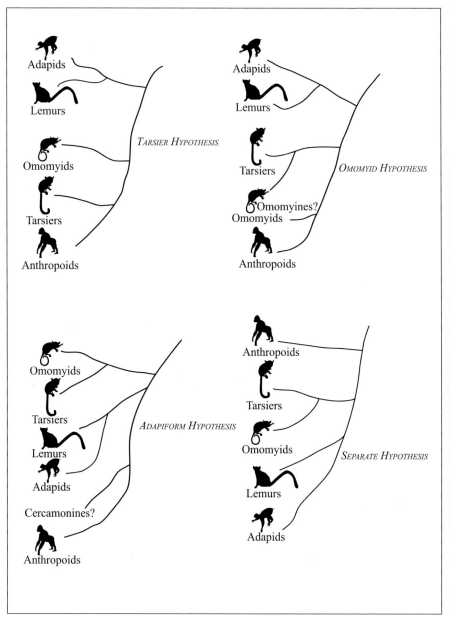

Four theories of primate evolution, before the Eosimias hypothesis.

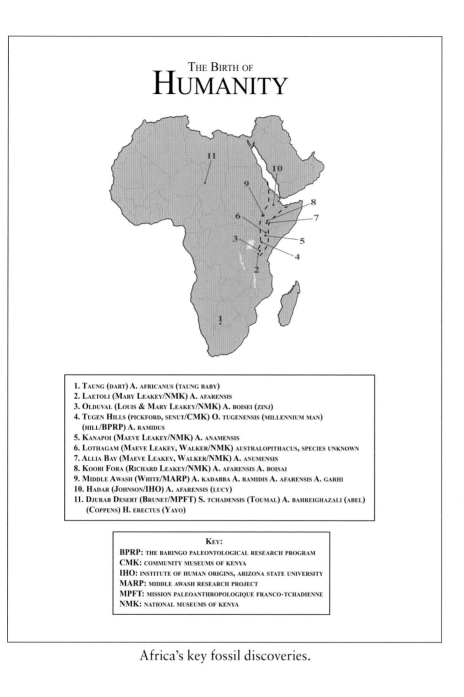

The Birth of
HUMANITY

1. **Taung (Dart) A.** africanus (taung baby)
2. **Laetoli (Mary Leakey/NMK) A.** afarensis
3. **Olduval (Louis & Mary Leakey/NMK) A.** boisei (zinj)
4. **Tugen Hills (Pickford, Senut/CMK) O.** tugenensis (millennium man)
   (Hill/**BPRP**) **A.** ramidus
5. **Kanapoi (Maeve Leakey/NMK) A.** anamensis
6. **Lothagam (Maeve Leakey, Walker/NMK)** australopithacus, species unknown
7. **Allia Bay (Maeve Leakey, Walker/NMK) A.** anumensis
8. **Koobi Fora (Richard Leakey/NMK) A.** afarensis **A.** boisai
9. **Middle Awash (White/MARP) A.** kadabba **A.** ramidis **A.** afarensis **A.** garhi
10. **Hadar (Johnson/IHO) A.** afarensis (lucy)
11. **Djurab Desert (Brunet/MPFT) S.** tchadensis (toumal) **A.** bahreighazali (abel)
    (Coppens) **H.** erectus (Yayo)

Key:

BPRP: the baringo paleontological research program
CMK: community museums of kenya
IHO: institute of human origins, arizona state university
MARP: middle awash research project
MPFT: mission paleoanthropologique franco-tchadienne
NMK: national museums of kenya

Africa's key fossil discoveries.

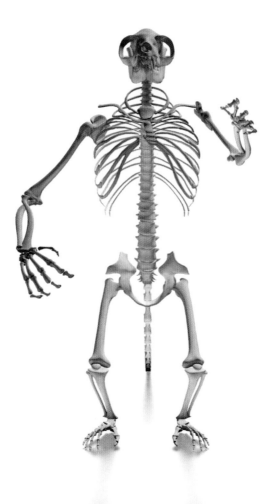

An early three-dimensional reconstruction of Ida's skeleton.

An early three-dimensional reconstruction of Ida's skeleton.

An early three-dimensional reconstruction of Ida's skeleton.

An early three-dimensional reconstruction of Ida's skull.

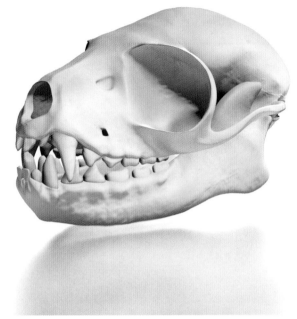

An early three-dimensional reconstruction of Ida's skull.

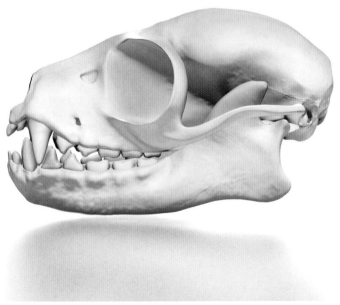

An early three-dimensional reconstruction of Ida's skull.

years ago (not long before, say, the founding of Jericho). Some assume she must be close to *Homo sapiens,* largely because of her recent date. But if she wasn't sick, she could be a late survivor of *Homo erectus,* closer to Peking Man than to modern people. This would be remarkable, though not quite so remarkable as might be supposed. Animals that are isolated on islands for long periods often evolve into miniature forms. There are various forms of dwarf elephant, not least on some islands of the Mediterranean. Sometimes too, island forms escape the extinctions that befall their grander relatives on the mainland. So it was that there might well have been dwarf mammoths on the island of Wrangel, off the coast of Siberia, as recently as four thousand years ago—contemporary with the pharaohs. More and more, as paleontology and archaeology progress, we find that prehistory and history overlap—and *Homo floresiensis* could be one such marvelously intriguing example.

The very first people who were more or less like ourselves—who would not have attracted too many stares on today's public transportation—date from about five hundred thousand years ago. The earlier types are commonly called *archaic,* and some of them are ascribed to their own species, such as *Homo heidelbergensis.* On the whole, though, the archaics are a confusing lot; probably different enough to look at but perhaps not different enough to prevent them from interbreeding if the different types came into contact with each other. Some biologists are content to define species according to appearance, but others say that different populations should be placed in different species only if they cannot readily hybridize; and some, pragmatically, say that different populations are different species if they look different and also, in practice, have nothing to do with one another, so their gene pools are isolated in practice even if in theory they might be combined by interbreeding.

The earliest of the decidedly robust people known as Neanderthals, *Homo neanderthalensis*, date from about two hundred fifty thousand years ago, much later than the first-known archaic *sapiens*. They lived from the Middle East up to northern Europe and they were still surviving in Gibraltar around twenty-eight thousand years ago. The Neanderthals have usually had bad press—commonly described as brutish and assumed to be stupid, and usually depicted even in modern museums as if they were dreadful scruffs, with masses of tangled black hair and unkempt beards. But the more that is discovered about them, the more sophisticated they appear to have been; for example, there is evidence of ritual burial and body painting. Since the most extreme types lived in the north, they were likely to have been blond or red-haired, like Scandinavians and Scots. And there is absolutely no reason to suppose that they were scruffy. The long hair of humans in general, and the beards of men, serve primarily as sexual attractants; and the whole point of such attractants is that they should be neat and tidy. You never see a wild animal that is scruffy unless it is sick or has been in a fight. Primates in general spend a great deal of their time grooming. This is a sign of health, which itself is a sexual attractant, and it implies freedom from parasites, which is vital for survival. You never see a scruffy human being, either, in a state of nature. Most of the world's surviving tribal people are immaculately turned out, and sometimes most elaborately. The idea that Neanderthals were scruffs is just another piece of Victorian and somewhat racist prejudice.

Because the first fully modern humans—the kinds known as Cro-Magnons—first arrived in Europe around forty thousand years ago, this means that they and the Neanderthals lived side by side for at least ten thousand years. How they got along is an intriguing question. Many have suggested that the Cro-Magnons were cleverer than the Neanderthals and simply hunted them. Others point out that there is no evidence of conflict in the archaeological record but suggest that the Cro-Magnons, being cleverer, simply outcompeted them. Others suggest that the two species (depending on how you define the word *species*) simply interbred, so that all of us now carry Neanderthal genes, some more than others. It's an intriguing notion. Indeed, just as the yeti could be a folk memory of *Gigantopithecus,* so the pagan gods of northern Europe—Thor and Wotan and the rest—could well be a folk memory of the last Neanderthals, with their great flaming manes of red or golden hair.

Sometime in the past forty thousand years, *Homo sapiens* invented farming. By about ten thousand years ago—the time of the Neolithic Revolution—agriculture was practiced on a scale large enough to show up in the archaeological record. The oldest-known cities date back about ten thousand years too, though again, there might have been much older ones. Writing is known from at least five thousand years ago.

After that, well, we are into modern times. We apply the epithet *ancient* to the people of Abraham and the Egyptians of the pharaohs and the Assyrians, the Greeks, and the Romans. But they were not ancient. They didn't have electricity, and horses and sailing ships were their only transportation, but apart from that, they were just like us. *Ancient* means something altogether different. Ancient, but just as relevant to our history, is Ida.

CHAPTER EIGHT

WHO AND WHAT IS IDA?

With the variety of primates that live and have lived on the Earth, and with incomplete yet stunning hominid fossil discoveries like Lucy, why does a nondescript animal somewhat like an oversize squirrel, squashed to the thickness of a German beer mat and mysteriously turned to stone, really matter?

Why does Jørn Hurum, who first acquired this petrified creature for the University of Oslo's Natural History Museum, say that it "will be the one pictured in the textbooks for the next hundred years"? Why does he suggest that scholars will flock to Oslo just to examine it? Why does he call it "a dream"? Why does he suggest—with some relish—that it will take him and his carefully chosen team of colleagues years to describe its anatomy in proper detail? And why does he think it is worth such a huge investment of time?

Hurum is not alone in his opinions. A colleague in this study, Jens Franzen, who has worked at the Senckenberg Museum in Frankfurt for the past forty years, calls this nondescript animal "the eighth wonder of the world." He thinks its impact on the world of paleontology will be "somewhat like an asteroid falling down to Earth."

Professor Philip Gingerich of the University of Michigan has been looking at fossils of many kinds since the 1970s, and for him, it is a Rosetta stone. The original Rosetta stone enabled scholars of the early nineteenth century to translate two unknown Egyptian-language scripts into Greek, a language they knew through and through, and hence to work out for the first time what the hieroglyphs meant and to read the hundreds of formerly cryptic messages on the walls of many tombs. For a slab of petrified bones, that's a lot to live up to.

So, what's so good about it? Everything, is the short answer. To begin with, it—*she*—is one of the finest fossils ever found, one on a very short list of all-time greats. She is extraordinarily old—about 47 million years old, from the early Middle Eocene. And for a creature of such antiquity, she is in a miraculous state of preservation. Her skeleton was not flattened by a truck but by a set of circumstances that can have occurred very few times in the life of the planet. The skeleton contained within the thin flat slab of shale is almost complete. Around the skeleton is the shadow of the fur; and among the bones, where the intestines would have been, are the fossilized remains of the creature's last meal. That such an object should have come into existence at all is remarkable beyond imagining. That it should have ended up in a museum where it is properly appreciated and can be studied at length by the world's leading experts for the rest of time seems like a miracle. But there it is, in Oslo.

As the icing on the cake, this creature is not of a kind that is peripheral to our lives. Emphatically, it is not a squirrel. It is

very definitely a primate—part of our own order, and potentially able to throw light on our own ancestry. As we have seen,
the fossil record has produced many thousands of fossil primates of all ages, now assigned to hundreds of different species
in several families. But they are all only partial, and some of
them are mere fragments—usually just bits of teeth, or teeth
and jaws. There is plenty of room for confusion and much that
is simply unknown. The information that does exist—on the
one hand plentiful and on the other hand so obviously full of
gaps—has led paleoprimatologists to propose four different
family trees. Two of them are somewhat similar, but the other
two are radically contrasting. This primate from Messel is
the *only* complete fossil primate of any age that has ever been
found—and with an outline of flesh and stomach contents too.
Moreover, it comes from a highly significant age—the time
when the anthropoids, our own subgroup of primates, must
have been arising from the prosimians.

Put everything together and this humble creature from the
Messel shale pit is remarkable indeed: uniquely and extraordinarily well preserved; from an era of particular significance
in primate history; and complete enough to help us pick our
way through the competing pictures of the primate family
tree. This creature can help tell us who we really are and
where we come from.

So let us look first at what these four competing family
trees entail.

The Family Tree of Primates

Within the order of the primates, there are six main lineages.
The most ancient, and presumably including the common

ancestor of all the rest, are the archaic primates. Next, reaching their peak in the Eocene, we have the adapiforms on the one hand and the omomyids on the other (where the omomyids are presumed to include the microchoerids). During the Eocene the three modern groups come on board. Two of them were still of prosimian grade: the strepsirhines, which include the lemurs, bush babies, lorises, and pottos; and the tarsiers. The remaining group, which qualifies both as a grade and as a clade, is our own: the anthropoids.

How are they all related to one another? And who is ancestral to whom? As ever, these questions are not quite as simple as they seem—and they call for a short diversion.

Ancestors and Sister Groups

The two concepts of relationship and ancestry are closely linked. Any group of creatures that is ancestral to another group must be related to that group, just as parents are related to their children. But of course this doesn't work the other way around. Just because two creatures are related does not mean that one must be ancestral to the other. They might simply be brothers and sisters or cousins. Clearly too, no group can be ancestral to another group unless it lived at an earlier time. But there again, creature A might be related to creature B and might have lived a long time before creature B, but creature A might not be an ancestor. Uncles and aunts are not directly ancestral to nephews and nieces.

There is one further complication. Suppose we suggest—as many biologists do—that the adapiforms were ancestral to the modern lemurs. This does not mean, of course, that all the adapiforms turned into lemurs, which form a true clade.

The assumption is that they all arose from just *one* small group of ancestors, which all belonged to just one species. So if we say that adapiforms gave rise to lemurs, we really mean—and can only mean—that *one* of the adapiforms was the lemur ancestor. But it is impossible to say which one; and since we know that the fossil record is far from complete, it's a fair bet that the precise group of adapiforms that included the mommy and daddy of all lemurs has been wiped from the record forever. Furthermore, we know that although all the adapiforms went extinct a long time ago, they did not all go extinct when the first lemurs arrived. While one small population of adapiforms was turning into lemurs, the rest carried on being adapiforms.

So the best we can say about all the adapiforms we know of is that they are very closely related to the lemurs, and that they share a common ancestor. That common ancestor is exclusive to the lemurs and the adapiforms, just as your parents are exclusive to you and your siblings. While we cannot claim that any particular adapiform we may happen to know about is unequivocally the ancestor of all lemurs, we can reasonably claim that all the adapiforms that we know about (and all the adapiforms that ever lived) at least have a kind of sibling relationship with the lemurs. So the adapiforms and the lemurs are said to be sister groups.

When we have two sister groups in which one of the sisters is far older than the other, there remains the possibility that the older group might and probably does contain the ancestor of the younger one. So, to be properly "parsimonious" (as the expression has it), taxonomists tend to present adapiforms and lemurs as sister groups, not as ancestors and descendants, but the implication is that the more senior of the so-called sister groups does in fact include the ancestor of the other.

Bearing in mind the conditional clauses and the caveats, who among the known groups of primates is presumed to be ancestral to whom? What does the family tree of primates look like? In the true spirit of paleontology and of taxonomy, nothing is straightforward. At present there are four leading hypotheses.

Four Scenarios

All of the four contending hypotheses begin with a few common assumptions of the kind that run through all biology. Scientists agree that the group we call the primates is indeed a true clade, meaning that all the creatures we recognize as primates, in all the six main lineages, sprang from a common ancestor who was exclusive to the primates (no squirrels, no raccoons, no pachyderms). This shared ancestor may or may not have been one of the archaic primates that we already know about. But for the purposes of argument, this does not matter; if the common ancestor wasn't a creature we already know about, then it's a fair bet that it was similar to the ones we do know. All the contending hypotheses assume too that each of the various subgroups is itself a clade. Most important, all assume that the anthropoids are a true clade—that the New World monkeys, the Old World monkeys, and the great apes all arose from the same prosimian ancestor. This may seem too obvious for words, but one very popular hypothesis from the past claimed that the New World monkeys arose independently of the two Old World groups—from a different prosimian ancestor—and that the similarities between New World and Old World monkeys were due to convergence. All of the current hypotheses reject that idea.

The four hypotheses are concerned with the whole family tree of primates: they want to show how all six recognized groups relate to one another. To keep the description simple, however, we will reduce this grand ambition to these questions: How are the anthropoids related to the other groups? Which of the other groups is the sister group of the anthropoids? Which of them could therefore contain our own ancestor?

One of the four hypotheses, the tarsier hypothesis, says that the anthropoids and the tarsiers are sister groups: that the anthropoid ancestor also gave rise to the tarsiers. The omomyid hypothesis says that anthropoids arose from the ranks of the omomyids. The adapiform hypothesis says that our ancestor lies somewhere among the adapids. The fourth, the Eosimias hypothesis, says that our ancestor was a monkey-like primate from the Eocene who lived in Asia. How do we test these hypotheses? How do we decide between them? One way is to find plausible common ancestors in the fossil record: creatures of the right age that *could* have given rise to both of the putative descendants. In all cases we have to ask, "How will we recognize the putative common ancestor when we see it? What kind of features must some ancient fossil possess to be a plausible candidate?"

First, a common ancestor candidate must have features in common with *both* of the supposed descendant groups. For example, the tarsier hypothesis says that tarsiers and anthropoids are sister groups. So any fossil that could pass muster as a common ancestor must possess features both of tarsiers and of anthropoids. If it is already committed to being a tarsier, with no obvious anthropoid features (or is already far more anthropoid-like than tarsier-like), then it cannot be the common ancestor of both. Neither, however, should it be too specialized in some other way. If we found some ancient creature that was a bit like a tarsier and a bit like an anthropoid but was also specialized to swing through the trees like a modern spider monkey, we would have to say that this creature could be related to the common ancestor of tarsiers and anthropoids but it obviously could not be the ancestor itself. It had already taken off on some evolutionary path that was leading in a quite different direction. In short, the candidate has to be sufficiently primitive to retain some flexibility and still be able, at least in principle, to evolve into either and both of the descendants. On the other hand, it mustn't be too primitive. For example, we could say that some archaic primate was theoretically able to evolve both into tarsiers and into anthropoids, but since its form was so generalized, it was also capable of evolving into lemurs. Very general features tell us nothing about special relationships.

So we are looking for a candidate ancestor that combines special features of both putative descendants; is not already too biased toward one kind or the other; is not specialized in ways that have nothing to do with either; and is not so primitive that it has the potential to become anything at all.

With all of that in mind, let us look a little more closely at the four competing ideas.

1. The Tarsier Hypothesis

The idea that tarsiers and anthropoids are closely related dates from the early twentieth century and is based on features of the living animals. Logic seems to suggest that anthropoids must have descended from some prosimian. Among living prosimians, the tarsiers tend to look much more anthropoid than the lemurs and the other strepsirhines do. In particular, as the British anatomist Reginald Innes Pocock pointed out in 1918, the upper lips and noses of anthropoids and tarsiers look very similar, whereas the upper lips and noses of lemurs and their relatives are very different. Anthropoids and tarsiers both have dry noses and an undivided upper lip. On the basis of this observation, the two groups could reasonably be combined into a bigger group termed the *haplorhines*, meaning "simple noses." Lemurs and other strepsirhines have a rhinarium—a wet tip of the nose—and a groove that runs from the rhinarium through the top lip to Jacobson's organ in the roof of the mouth. This is implied by the term *strepsirhines*, which means "complicated noses."

Moreover, although tarsiers are nocturnal and have the big eyes that go with the nightlife, they lack a tapetum. This is the layer of reflective cells in front of the retina that is a primitive feature of mammals—it's what causes the eyes of most kinds to shine when you flash a light on them, as with cats. Anthropoids lack a tapetum: the red eye that results in flash photographs of humans is caused by light reflected from the blood in the retina itself, not from a tapetum. Tarsiers would surely benefit from a tapetum. So why don't they have one? Perhaps, the tarsier hypothesis says, because the common ancestor of the tarsiers and the anthropoids had already lost the tapetum before the two groups diverged. On another shared, minute point of detail, only a very few primates are unable to synthesize their own vitamin C; these include tarsiers (although they are strict carnivores—mainly insect eaters) and human beings. Finally, and more convincingly, we see similarities in the skeleton. Most obviously, the orbit of the tarsier forms an almost complete bony socket, much resembling that in anthropoids. It doesn't just make do with an orbital bar, like the other prosimians.

So in this scenario, the anthropoids and the tarsiers are seen as sister groups; the two together are then seen as sisters to the omomyids; and all the living strepsirhines, together with the adapiforms, form a quite separate primate branch.

This hypothesis is neat and plausible, but most biologists now feel that it is incorrect. The similarities of lips and noses seem merely superficial. The wet nose is presumably the primitive condition—the very early archaics presumably had wet noses—but the tarsiers and anthropoids probably evolved their dry noses independently. The haplorhines, therefore, are not a true clade. They include two different lineages of primates that look similar because of convergence. If we look closely at the eye sockets too, we find that although they appear similar in tarsiers and anthropoids, they are constructed in a quite different way, incorporating elements from different bones. So here we have an instance of convergence yet again.

2. The Omomyid Hypothesis

This idea looks somewhat similar to the tarsier hypothesis, but it is significantly different. It starts with the idea that the living tarsiers have much in common with the ancient omomyids. Indeed, it posits that one or another of the omomyids was the probable ancestor of the tarsiers. So, the idea has it, the tarsiers and omomyids are sister groups and can be combined into one group. Then this group would be the sister group of the anthropoids, meaning that the anthropoids share a common ancestor with the tarsiers-plus-omomyids. As in the tarsier hypothesis, the adapiforms and the strepsirhines are seen to occupy a quite different branch of primates.

But, say Philip Gingerich and his associates, "The omomyid fossils known from North America and Europe disclose no special relationship with anthropoids." Instead, some of them at least look as if they are becoming specifically tarsier-like. Some have very large orbits; and in *Necrolemur,* one of the European omomyids (first found in Quercy, France), the shinbones (tibia and fibula) are beginning to fuse, and the foot bones (tarsals) are becoming very long. A creature that is already apparently on the way to being a specialist tarsier is not likely to have given rise to the anthropoids, which have a very different set of specialties.

3. The Adapiform Hypothesis

This is the idea preferred and championed by Gingerich. It says that the anthropoids belong on a branch that includes the adapiforms and the strepsirhines. So, in absolute contrast to the first two hypotheses, the anthropoids are seen to be quite separate from the omomyids and tarsiers. This time, it's the omomyids and the tarsiers that are out on their own.

The adapiforms, however, are a complicated lot. As we have seen, the lemurs are clearly close to the adapids. The anthropoids have more in common with the notharctids—the primarily North American versions of the adapiforms. So this idea is radical; we don't usually think of lemurs as our cousins. At best, we feel they are our second cousins. But quite a few elements point to this hypothesis. The lower incisors of adapiforms (though not generally of lemurs) are far more like those of anthropoids than those of omomyids. In adapiforms, the lower incisors are small, stand vertically, and are spatulate (shovel-shaped)—just like ours. The lower incisors in omomyids tend to be larger and pointed and protrude forward. In several lineages of adapiforms, including *Notharctus,* the lower jaw is composed of only one big bone, as in anthropoids; it is a fused mandible, with no division into two bones and with a join at the chin. No known omomyid has a fused mandible. Adapiforms tend to have fairly large canines, as monkeys and apes do. But in omomyids, the canines are reduced. (To be sure, human beings have reduced canines too, but the fossils tell us that our more recent ancestors, from the Pliocene, had bigger canines, more like an ape, so our small canines are a late development.) It also seems that some adapiforms, again including *Notharctus,* show some sexual dimorphism, again typical of many monkeys and some apes (although, of course, not all). On a point of detail—the kind that often betrays true relationships—many adapiforms have lost a particular cusp on the lower molars, known as the *paraconid* (which is present in the most primitive primates). And so have anthropoids.

The similarities do not stop there. The ankles of adapiforms are more similar to those of anthropoids than omomyids' are. In omomyids, the heel bone is elongated, and in adapiforms the heel bone is more heel-like; present just to turn the corner from the leg into the foot, as in ourselves. So the omomyids and their European cousins, the microchoerids, are too specialized to be ancestors of the anthropoids. As we noted earlier, they seem already to be moving in the direction of the tarsiers, with their large eye sockets, their forward-pointing teeth, and their already elongated ankle bones.

The skeletons tell all, for when we look at living tarsiers and living lemurs, the tarsiers seem to look more like the anthropoids, with their dry noses and unsplit lips. Since the tarsiers are clearly related to the omomyids, and the lemurs are clearly related to the adapiforms, it seems at first glance that the ancestors of the anthropoids should be shared with the omomyids rather than with the adapiforms. But when we look more closely at the fossils, as Gingerich argues, we see that the adapiforms—or at least some of them—are closer to the anthropoids, and more likely to include the anthropoids' ancestors.

4. The Eosimias Hypothesis

The fourth idea says that all of the above are wrong. This hypothesis, championed in particular by Chris Beard of the Carnegie Museum of Natural History in Pittsburgh, says that the true origins of the anthropoids lie in Asia. Beard puts particular store in a tiny fossil primate about the size of a modern pygmy marmoset that was found in China in 1994 and dates from the early Middle Eocene, about 45 million years ago—just a bit younger than Ida. This creature, Beard says, shows a "unique combination of primitive and advanced anatomical features." Indeed, it is, he says, a "primitive anthropoid." This notion is reflected in its name: *Eosimias,* meaning "Dawn Monkey."

As with all of the hypotheses, there are criticisms of the *Eosimias* story. One has to do with the fossil itself, which some say is simply not enough like an anthropoid to justify such enthusiasm. Others point out that if the *Eosimias* hypothesis is true, then this means that most of the fossils we would need to tell the complete story of anthropoid origins are missing. We have a great many adapiform fossils by which to explore the adapiform hypothesis, and a great many omomyid hypotheses to support—or to throw doubt on—the omomyid and the tarsier hypotheses. But if *Eosimias* is indeed the first anthropoid (or very like it), where are its precursors? And who were they? An obvious reply to this is that the fossil record is notoriously spotty—a few bits here and a few bits there—and we should not be at all surprised if most of what we need to know has not been found yet or perhaps no longer exists. Besides, most of the studies so far have been done in Europe and North America, because Europeans and North Americans historically have been the most involved, and they tend to look first in their own backyards. Asia in general and China in particular have only recently come seriously online. China has already yielded some fabulous finds in many areas (including the famous feathered dinosaurs), and one could say, we ain't seen nothing yet. On the other hand, if we are going to invoke hypothetical lineages of creatures of which we as yet have no intimation, where does that stop?

How can we resolve such an issue? In truth, we cannot—not definitively. The movie of the deep past cannot be rerun, and nature always springs surprises. But any hypothesis that says animal A is the ancestor of animal B is greatly strengthened if fossils can be found of a transitional kind—ones that combine features of the two and show how A might have evolved into B. These desired transitional types are often missing and are popularly known as "missing links." All fossils are elusive, and missing links are likely to be more elusive than most.

Missing Links and Ida

When Charles Darwin first proposed publicly and formally in 1859 that evolution was the way of the world, not everyone was convinced. Among the unconverted were some of the leading biologists of the time, including Darwin's archrival Richard Owen. Owen was perhaps the leading anatomist and paleontologist of his day, certainly in Britain, just as Georges Cuvier had been in Europe some decades earlier. Owen said that if creature A had indeed evolved into creature B, then the fossil record ought to contain creatures that were half A and half B. But on the whole it didn't.

Owen suffered a setback and Darwin's star rose significantly when *Archaeopteryx* turned up just a few years later. Owen had asked in particular, if birds evolved from reptiles, as Darwin claimed, where was the half-reptile, half-bird? Then the very creature turned up at Solnhofen, in Germany. (Germany seems to be emerging as a key provider of missing links.)

As we discussed in chapter 6, statistics conspire to ensure that missing links are especially rare. Fossilization in general is a rare event, and the animals most likely to be fossilized are the ones that are common and widespread—widespread so that at least some of them live in those special places where fossilization is more likely to happen. But when new kinds of creatures arise—the first birds, the first monkeys, the first human beings—there must obviously be only a few of them, in one particular place, which may or may not be a place where fossils are likely to form. New life forms that find themselves in new ecological niches tend to evolve rapidly too, as the first human beings did. So the first ones on the scene quickly turn into something else. The less time any one kind of creature spends on Earth, the less likely it is that any individuals will fossilize.

Genuine transitional types—missing links that are no longer missing—are very rare, very valuable, and much prized. To be convincing, they must combine features of both the supposed ancestor and the supposed descendant. They must not be too specialist in their own ways, with special features that are just not present in the descendants. But they must not be too primitive either, where the only features they have in common with the supposed descendant are the kind that all vaguely related creatures are likely to share. A fossil that offers such insights must be of very high quality indeed. Most fossils, after all, are just fragments; in the case of mammals, including many species of extinct primates, they are bits of teeth and jaws. It is very hard to demonstrate that a small piece of some ancient animal has a combination of features that link some putative ancestral type to some putative descendant.

But Ida seems to fit all the required bills. She is of extraordinary quality, containing a great deal of what we need to know. And she does seem to be transitional, poised between one of the ancient types and the anthropoids. This is why she has invoked such enthusiasm.

So where does she fit in the evolutionary tree of primates? Which, if any, of the current hypotheses does she support? It is time to look at her in detail.

Ida

To cut to the chase, Ida reinforces Gingerich's idea of primate evolution that anthropoids descended from adapiforms, although a very particular group of them. The tarsiers, so long thought to be the closest living relatives of anthropoids, now are thought to belong on a quite different branch; and so are the omomyids, which are clearly related to the tarsiers. Gingerich rejects the *Eosimias* idea of human ancestry largely because, he says, the fossil evidence is too sparse and too uncertain. But Ida seems to show just what is needed: transition from adapiform to anthropoid. She has all the right qualifications. She has some features in common with lemurs, but none of the extreme specializations of modern lemurs, which clearly mark a lemur as a lemur. She has some features that we associate with anthropoids, yet she is not a full-blown anthropoid. In short, she provides hints of what might become but is primitive enough to be flexible, not yet committed to one particular lineage. She could in principle have given rise to both lineages. She is also precisely the right age: just the age you would expect a prosimian-anthropoid link to be.

Overall, she is the size of a large squirrel; or, in primate terms, the same size as an Eastern woolly lemur. From the tip of her snout to the end of her tail she is about one foot, ten inches (57 centimeters) long, but she was not full-grown when she died and presumably would have grown to more than two feet (60 centimeters). More than half of her length was tail, however—a tail with thirty-one vertebrae. As an adult, she probably would have weighed about the same as an Eastern woolly lemur: almost three pounds (up to 1.3 kilograms)—although that would be a big one. She was a substantial creature, in short.

She had a wide-eyed, monkey-like face. The orbits, and therefore her eyes, are big, meaning she was probably nocturnal. She had a high forehead and a fairly rounded cranium, with short ears hidden by the fur; these are revealed by the miraculously preserved "skin shadow." She didn't have a crest of bone along the top of the skull for extra-powerful chewing muscles, but she did have a horizontal crest of bone along the back of the skull (known as the *nuchal crest*), for muscle attachment.

Her teeth tell us all kinds of things. In general, as with all parts of a fossil, the functional bits of the teeth tell us about her lifestyle—specifically, what she ate. The functional bits of creatures are subject to heavy selection pressure and must adapt rapidly to way of life. But it's the bits that seem to be nonfunctional—the particular layout of cusps on the molars, for instance—that reveal her true relationships. The nonfunctional details tend to remain unchanged from generation to generation, since there is no reason for them to change. The details in general excite no interest among the uninitiated, but they are just what the specialist is looking for.

So what can we make of Ida's teeth? First, they tell us her age. Her canine teeth are still milk teeth, although X-rays show the permanent teeth well formed behind them, and they are big, just like those of anthropoids. Most of her molars have not yet erupted, so she was a juvenile—no longer a young child but not yet an adult. The Oslo team suggests that she was around six months old when she died.

What was her teeth's function? What did she eat? Her molars in general form are like ours, with small, rounded cusps and deep basins in between—ideal for fruit, it seems. She is the right size for a fruit eater. She would have needed a fairly rich diet but would not have needed to eat a high proportion of insects, as a smaller creature would have had to do. We would expect her to have eaten some insects, yet her gut contents tell us a somewhat different story. But we will come to that.

Her incisors are compatible both with adapiform and anthropoid connections but are most unlike an omomyid's. They are fairly small and set vertically. The lower incisors of omomyids are bigger and point forward. But this fossil clearly was not a lemur, whose lower teeth also point forward and are pointed but form a toothcomb, for grooming.

Most intriguing of all are her premolars. She has only two premolars on each side on the upper and the lower jaw. In front of them a remnant of a milk molar is still present, which is reduced to a small peg. Virtually all other primates have three premolars on each side, both upper and lower. The only exceptions are the Old World monkeys and apes and human beings, which have only two premolars per side per jaw. The loss of one or more premolars from the primitive condition is called *antemolar reduction*. To Gingerich, the antemolar reduction of Ida is highly significant.

The same principles apply to an evaluation of the rest of the skeleton: whatever is obviously functional tends to reflect way of life, and it is the points of detail that suggest true relationship.

Ida had shortish arms and much longer hind legs—the classic shape of a clinger and leaper. Tarsiers are clingers and leapers, and so are many strepsirhines. But the heel bone in Ida was anthropoid-like; it served just to turn the corner between foot and leg. It was not elongated to virtually provide another length of leg, as in tarsiers.

Ida's hand is stout, to a much greater degree than in creatures of the same age that are thought to be related: *Europolemur, Godinotia,* and *Notharctus.* The thumb is small and diverges almost at right angles to the rest of the hand. This is like a lemur's but not like an anthropoid's. The ends of the fingers are scutiform—shield-shaped—suggesting that Ida had nails, not claws. As in a modern lemur, the fingers (apart from the thumb) are very long compared to the metacarpals, the bones of the palm of the hand.

Ida's feet are like her hands. The strongest of the tarsals—foot bones—by far is the first, the one that leads to the big toe: it is very big, and, like the thumb, it stretches out at right angles to the rest of the foot. Its saddle-shaped joint shows that it was opposable. Again, all the toes clearly had nails rather than claws.

In the feet too we find a feature that is very un-lemur-like; or, rather, we find the absence of a feature that is very characteristically lemur-like. The lemurs and their kind have a grooming claw, or toilet claw, on the second toe (the toe next to the big toe, corresponding to the index finger): a blunt, hooked, narrow structure used to groom the fur. But such a structure is not identifiable in Ida. Clearly, although she has much in common with lemurs, she is not herself a lemur.

So, to the very big questions. To whom is Ida related? Where does she sit on the grand family tree of primates? Whose ancestor was she, if anybody's? Could she conceivably have been ours? Should we call her our grandmother?

To Whom Is Ida Related?

In many ways, Ida resembles some of the other primates already known from the Eocene. The overall shape of her skull is similar to that of *Mahgarita stevensi*, which was found in Texas. Her skull also has a lot in common with *Pronyctice-bus gaudryi*, although Ida is more robust. When Franzen first described Ida's slab B—what should have been the matching half of the Ida fossil—he assigned it to the species *Godinotia neglecta,* which was already known from another Eocene site in Germany, at Geiseltal. But Ida has now been examined in many ways, including X-ray and microtomography, and she is clearly different from *Godinotia,* especially in the proportion of the limbs. Her limbs are different too from the two species of *Europolemur* that are known from Messel, and from the North American *Notharctus,* and she differs from *Cercamonius* from Quercy in details of the cusps of the teeth.

Putting everything together, the team that gathered at Oslo has concluded that Ida belongs to the adapiform family known as the Notharctidae. More particularly, she belongs to the subfamily Cercamoniinae.

But she is definitely a new species, different from any of the cercamoniines known already. The team will announce Ida's Latin name when they publish their findings, and they intend to dedicate her to her location and to Darwin, without whom our understanding of evolution wouldn't exist and her importance would be obsolete.

Overall, says the Oslo team, *Ida* is in general lemur-like and yet "represents rather primitive primates, foreshadowing developments in an anthropoid direction." To summarize: Ida combines features of a lemur-like kind with features of an anthropoid kind—and, crucially, she is not committed to either course.

Who Was Ida and How Did She Live?

Ida's teeth suggest that she was mainly a fruit eater, and we might suppose, by comparison with modern primates of similar size and general form, that she supplemented her diet with insects. But such is the wonder of Messel, we need not live with mere suggestions. The contents of her gut are conserved. Her last meal.

In the spirit of a forensic pathologist, Franzen examined Ida's gut content, and he found among the debris a strange "scute"—a shieldlike scale. At first he assumed it was the scale of a fish—not a fish that Ida had eaten but one that had fallen into her remains as she fossilized. "This is quite com-mon, to find isolated scales of fish," he says.

But, says Franzen, "I was looking again at that scale, and suddenly the scales fell from my eyes—because I recognized the inner structure of that scale was not the structure of a fish scale, but there were cell walls typical of plants or, more precisely, of a seed." So then Franzen summoned a colleague who specialized in fossil plants, "and he stepped in and looked quite skeptical and then when he looked through the microscope, the scales fell from his eyes also."

The two scientists next looked at the gut content under a fluorescent microscope, which picked out all the other frag-ments of leaves in the fossil, and then applied a scanning elec-tron microscope, which offers astonishingly fine resolution in three dimensions, and picked out all the details of the seed. But although it seems likely that Ida would have eaten insects that came her way, they assumed, they found no sign of any. This wasn't an accident of preservation—there are plenty of bits of insects in the gut contents of other mammals from Messel. In fact, says Franzen, "We really could conclude that this animal was only eating leaves and sometimes fruit."

From the general shape of her skeleton and her grasping hands and feet, we can see that she was arboreal. But we can say more than that. Her hind legs are not as long, relatively speaking, as those of a modern-day indri or a bush baby. But they are longer than her arms; and the outline of flesh around her thighs shows that they were very muscular. So, like indris and bush babies, she was a clinger and leaper; clinging to vertical trunks and then taking off in one great jump several yards to the next one, high above the forest floor.

From one specimen we cannot, alas, infer much about a social life. If we had males and females, we could discover whether or not there was sexual dimorphism in Ida. If there was, then we would know that Ida was destined to live in a harem. If not, then she might well have grown up to be one of a monogamous pair. We do know that fruit eaters are sometimes rather solitary, like orangutans, but they also often go around in great mobs, apparently because fruit can be hard to find, and many pairs of eyes and noses are better than one. But to truly pin down the social life of an extinct animal, we need more than one specimen and examples of both sexes.

Why throughout this account have we been assuming that this ancient primate was female? Why call her Ida? Couldn't she be a Fritz or a Hans? Though we can often tell the sex of a fossil mammal from the shape of its pelvis, Ida died when she was too young for her bones to have been shaped for childbearing, so that does not help us. Ida is judged to be female not by what she has, but by what she has not. In most mammals, and in all primates except human beings, the penis is reinforced with a bone known as *os penis* or a baculum. Thus equipped, males of creatures such as bush babies go through life with a permanent erection. The baculum varies in shape from species to species, and externally the organ varies even more. In some it is club-shaped. Some have backward-pointing spines toward the tip. Some have spines at the back. As is common in insects, it seems as if the male organ is structured to fit in the female precisely, like a key in a lock. In some species, the penis serves as an organ of display, in courtship. In some the *os penis* is relatively enormous. If the smallest known species of bush baby were scaled up to the size of a human, its *os penis* would be about a foot long. Sometimes, with a fossil, the only bone of any kind found is an *os penis*. If such a bone had been present in Ida's skeleton, it would still be there. But it is not. So she must have been female.

Why, you might wonder, don't humans have an *os penis?* Perhaps, some biologists suggest, it is because men are too macho. All secondary sex characteristics in the males (or generally the males) of all animals are thought to be symbols of their vigor. Stags cannot grow huge antlers and roar vigorously and continually unless they are themselves big and healthy (and, in the case of stags, unless they are several years old and hence demonstrate their ability to survive). Peacocks cannot maintain and carry around their huge and functionally absurd tails unless they are free of parasites and feed well. Indeed it is *because* the antlers of the stag and the train of the peacock are such encumbrances that they reveal the vigor of their possessors. They demonstrate the "handicap" principle, which says that males of all kinds often handicap themselves specifically to show that they find life so easy that they can afford to be extravagant. For animals without an *os penis,* the only way to achieve a convincing erection is by hydrostatic pressure—which only a vigorous individual can generate. An erection achieved without obvious support may not tell a potential mate all she might want to know about her suitor's health, but it does tell her that he is not actually ill, at least in one important respect.

How Did Ida Die?

There is one last mystery to be solved. Why did Ida die before her time? And why aren't there hundreds of fossil primates at Messel? If one young individual could drown in the lake, why didn't more?

We shouldn't be surprised to find a lot of fish in the fossil remains of a lake, and the abundance of birds and bats at Messel is certainly striking. But all the years of exploration there have produced very few primates—just eight specimens, from three species: two kinds of *Europolemur* and now the complete *Ida*. Yet the lake was surrounded by tropical or semitropical forest, and surely it abounded with primates. How is it that so few of them perished in the lake?

The answer, perhaps, lies in Ida's left hand. As Franzen discovered when scrutinizing the skeleton in Oslo, it is broken. That is, she broke it in life, and at the time of her death, it was healing. Perhaps she found it hard to live up in the trees at around that point. Perhaps for a time she had to live on the ground. So she came to the lake to drink, was gassed by the CO_2, and fell in. As Jørn Hurum says, "Poor little girl!" Bats and some kinds of birds were overcome too because they flew low over the lake to look for insects, or to take a sip of water. Flying insects perished in the same way. But the primates stayed up in the trees. Water collects in tropical trees, in hollows and in whorls of leaves, so creatures that lived in the canopy had no need to come to the ground—unless their circumstances forced them, as Ida's did.

From her flattened skeleton, we can write Ida's life history, and attest that it was a short one. She was roughly like a midsize lemur in general size and shape, but was she literally a lemur, or was she some other kind of primate? If another kind, did she belong to a family that still exists, or was she of a type that has long since gone? Is she ancestral to any modern kinds? Is she, indeed, *our* ancestor?

Is Ida Really Our Grandmother?

While Ida was definitely prosimian, she had features that strongly suggest anthropoid leanings. She is also exactly the right age to be the long-sought "missing link"—no longer missing. But many paleoprimatologists have imagined that the true ancestor of the anthropoids, our own ancestor, must have lived in Eocene Africa. Certainly, the earliest anthropoid fossils are African: they come from the Fayum Depression in Egypt, about thirty-seven miles (60 kilometers) south of the pyramids. And though Messel was farther south in the Eocene, it was never part of Africa. Many might feel that this alone is enough to suggest that Ida was off the main chain, and that our own ancestor was living at around the same time, looking similar and living in the same way—just somewhere else. While those who suggest that would not call her Grandma, even they could well call her Aunt. And it's not bad at all to track down an aunt after an absence of 47 million years, especially when no such aunts have been tracked down before.

But there is also the fact that with the fossil record we have, and the inherent, absolute difficulty in defining the number of missing links in that chain, Ida is without debate a spectacular discovery—a link to primates past and, perhaps, as close a relative as we might imagine.

CHAPTER NINE

REVEALING IDA TO
THE WORLD

At its core, the narrative of human evolution is an evolving one, leading from how we came into existence to who we are. The details that we have reveal that all humans are interconnected and related at some root level. But the story is a frustrating and exhilarating work in progress, told in long chapters filled with many blank pages.

Much of the evidence comes from fossils, at best an imperfect means of speaking to us. For a fossil to form, the creature must die in specific conditions. It must be covered soon after death in order to prevent its being eaten by other creatures or decomposing due to the presence of bacteria. The creature's final resting place must remain geologically stable for millions of years. And then it has to be found, properly extracted, and compared to the rest of the record. Only at that point can a new piece be added to the puzzle of our past and, possibly, contribute to what we understand about ourselves.

Jørn Hurum and his team of scientists concluded from their

first comprehensive study of Ida that she will have a significant place in the story of human evolution. "This fossil will be the image of our early evolution for generations to come," Hurum predicts. "It is a symbol to every human being on the planet, no matter what age, race, or creed—we all share the same ancestors and cousins, we are all primates."

At her simplest, Ida was a small leaping primate, probably nocturnal and probably a vegetarian. She lived 47 million years ago in a tropical forest not unlike the ones found in South America today. She died after encountering toxic gas released from a crater lake, and a recently broken right wrist and weak left arm probably prevented her from scurrying to safety. Her remains were fossilized along with the gut contents of her last meal intact.

Remarkably, Ida displays characteristics of wet-nosed and dry-nosed primates, the prosimians and the anthropoids. However, she is a sui generis species. Her eyes are completely stereoscopic. She has a fused cranial plate, indicating increased brain growth. Her lower mandible is fused and her teeth are spatulate. Most important for placing her in the primate evolutionary chain, she doesn't possess a toothcomb or a grooming claw, which are innate lemur traits.

Hurum and his team have concluded that Ida documents the moment when the early primates were just about to split into two very different lineages. Each lineage was successful in its own right, but at the end of the anthropoid lineage is the human, the most successful primate to have walked the Earth. In other words, Ida appears to be an in-between species, or one of the long-sought missing links in evolution. Without a species quite like her, there would have been no modern-day lemurs, monkeys, or apes. She is, quite possibly, the biggest breakthrough in our understanding of primate evolution in more than thirty years.

What is certain is that Ida will open up the field of primate study, says Messel Pit fossil expert Jens Franzen. "On one side, it's for the first time clearly visible, even for laymen, what we are dealing with. There is a complete body [with a] soft body outline, and it's so wonderful, it's a miracle. On the other side, it will become a hot spot of debate. Just having such fossil evidence at hand, we are dealing with very solid ground, and so we are optimistically looking forward for that debate to begin."

As a nod to her broad significance, Philip Gingerich, the team's fossil primate expert, selected the scientific name for Ida, naming her genus after Charles Darwin because Ida would be introduced to the world on the two hundredth anniversary of Darwin's birth and the one hundred fiftieth anniversary of the publication of *On the Origin of Species*. The species name was derived from the first written account in 880 of the Messel Pit, where Ida was discovered.

In March of 2008, Hurum was wrestling with how and when he would present Ida to the world. Even at that point, he and his scientific team of Jens Franzen, Jörg Habersetzer, Wighart von Koenigswald, Philip Gingerich, and Holly Smith had collectively concluded that Ida was the most complete fossil primate ever found, and Hurum wanted to find a way to present their findings to the scientific community and the public in order to tell Ida's story in detail to the greatest number of people.

"I knew that a fossil can be metaphorically destroyed if you make a bad decision on who is to produce a documentary, and if it is explained wrong or ridiculously in the media," Hurum says.

Hurum had already been talking to documentary producer Anthony Geffen about doing a film on his discoveries in the Arctic archipelago of Svalbard. Geffen's company, the London-based Atlantic Productions, had produced several landmark documentaries. In April of 2008, Hurum invited Geffen, who began his career at the BBC, to Oslo to spend the day at the museum and discuss Hurum's next mission to the Arctic, where he was searching for fossilized sea monsters. The two spent five hours together in the museum, talking about their work, and a mutual respect developed.

Hurum then took Geffen to a late lunch at his favorite restaurant, a decidedly local joint with the best Turkish food in Oslo. He spent the meal questioning Geffen about how he maintained secrecy while making and marketing films about ancient discoveries. Hurum was somewhat nervous to show Ida to Geffen, because so few people had seen her picture or her remains. He wasn't even sure where to begin the story.

When Hurum pulled out his phone and opened a photo, Geffen leaned in and looked at the image of Ida. Knowing Hurum's work, Geffen immediately sensed he was looking at an important discovery. But he was utterly perplexed, and his first question was, "What is that?"

"This could be our oldest ancestor, and possibly one of the missing links," Hurum said calmly.

Geffen's first thought was that the creature looked like a cross between a monkey, the film character E.T., and a small human being. He studied the curious image. "Can I see it in person?" Geffen asked.

Hurum and Geffen returned to the museum. After Hurum

led Geffen through a long corridor in the basement and past security guards to the room where Ida was held, he unlocked a series of doors by swiping different key cards, and they finally reached a wooden door. The number 24 was above the door frame.

Hurum pulled out one final key card and opened the door. The room had several metal cabinets against the walls and a large table in the middle. Hurum walked over to one of the cabinets and unlocked it. He slid out a drawer, removed the fossil plate that was the size of a large serving platter, and placed it on the table. He switched on an examination light.

Geffen was stunned. "It just takes your breath away," the producer recalls.

Hurum stood there, feeling proud and somewhat relieved that at last he'd been able to share his secret and that the planning to show this specimen that had been unearthed twenty-five years ago could begin.

The Oslo Natural History Museum was about to be hit with the international spotlight. Hurum and museum director Elen Roaldset knew their credibility and that of the scientific team that examined Ida would be on the line.

"Anthony had worked with museums on discoveries," Hurum says. "We had a real personal connection about the goals we shared for discoveries. With all his knowledge and the films he had made, I thought he would be an ideal partner."

Hurum and Roaldset held several meetings with Geffen to discuss how to maintain the secrecy of Ida and how she would be revealed. The scientific paper would be the backbone of Ida's unveiling and the starting point for all future analysis by other scientists. Each scientist on Hurum's team contributed to the study largely in his or her area of expertise. Habersetzer wrote about carrying out the radiographic and micro-CT investigations and produced the plates of X-rays and three-dimensional models. Franzen, Hurum, and Gingerich described, compared, and explained the specimen. Hurum prepared most of the drawings, as well as casts and a reconstruction in conjunction with an artist. Franzen, von Koenigswald, and Smith investigated and described aspects of biostratinomy, the description of the process between death and burial. Smith, Gingerich, and Franzen solved the intricate interpretation of deciduous and permanent teeth. Prior to publication, their paper would undergo an independent scientific peer review.

Their paper would be the first official description of the 47-million-year-old primate. Since 1994, Hurum had published twenty peer-reviewed scientific papers, thirteen on Mesozoic mammals and seven on dinosaurs. But he knew that this one would easily be the most scrutinized of his career.

"The most important thing is that we put out a scientific paper that can withstand all the criticism it will get," Hurum says. "A lot of the thought behind assembling the dream team was to cover the basics of the Eocene and primate evolution in the Eocene."

While the paper was being written and edited, Geffen and his filmmaking team began making the documentary. The hope of the entire team was that Ida would be accessible for everyone—from children and laypeople interested in the ancestry of modern man to scientists poring over potential missing links in the evolutionary chain. For this reason,

Hurum was delighted that Sir David Attenborough joined the documentary project, noting, "I was extremely honored that our work had got the attention of the most important voice and face of natural history." When Attenborough saw the image of Ida, he was excited by the discovery and called it an extraordinary fossil.

"What we have here is like the Rosetta stone," Hurum says, referring to the Egyptian tablet that helped clarify scholars' understanding of hieroglyphic writing. "For the first time, you could look at hieroglyphics and decode them. It will be the same, because we will for the first time see what the characteristics in an early primate really are. Instead of guessing from single teeth and broken bones, we can start to understand them in the context of all the other science from around the world, because all the characteristics are present in one specimen."

The attention placed on and scrutiny of modern discoveries that potentially fill in the tree of primate evolution can be intense, and academic and journalistic debate continues on recent finds that garnered significant attention in the media, such as *Sahelanthropus,* dubbed Toumai, and *Homo flor.- siensis,* nicknamed the Hobbit. *Sahelanthropus* was a 6- to 7-million-year-old fossil ape found in Chad between July 2001 and March 2002, and the scientists who discovered the skull, led by Michel Brunet of the University of Poitiers in France, concluded that it belonged to the oldest-known ancestor of humans after the split of our branch of the primate tree from that of chimpanzees. Brunet and his team believed that this creature had been the first to stand upright and walk, leaving chimpanzees and apes behind. A nearly complete cranium, several pieces of jaw, and many isolated teeth were found, but numerous prominent dissenters, including University of Michigan anthropologist Milford Wolpoff, claimed that *Sahelanthropus* was merely an ape.

Homo floresiensis was an eighteen-thousand-year-old skull and parts of several bones that were discovered in 2003 in a cave on the Indonesian island of Flores. Anthropologists Peter Brown and Michael Morwood, who made the discovery, argued in a 2004 scientific paper that "the Hobbit," which stood just three feet (1 meter) tall fully grown and had a brain less than a third the size of a human's, was from a previously unknown species of hominid. Lee Berger, a paleoanthropologist at the University of Witwatersrand in Johannesburg, and his colleagues, however, claimed that *Homo floresiensis* was not a distinct species but rather a group of humans afflicted with a disorder that made their bodies unusually small.

While much of science is based on hypothesis and inspires vitriolic debate, like the ones that continue over *Sahelanthropus* and *Homo floresiensis,* the scientists who have studied Ida believe she could be a breakthrough on the level of Lucy or the Turkana Boy, the nearly complete 1.5-million-year-old skeleton of an eleven- or twelve-year-old hominid boy discovered in 1984. Whereas Lucy had proved that just over 3 million years ago, apes began walking upright on two legs, Turkana Boy was the most complete skeleton of a hominid since Lucy, and it has helped scientists with their study of body proportions of the first early humans.

Ida's completeness and age distinguish her. "Between Lucy and Ida, there are only enough specimens to fill a small room, and they are mostly incomplete," Hurum points out. Ida is more than 43 million years older than Lucy, a time frame that's almost impossible to comprehend. From Ida's age alone, she may open up one of the first chapters of what became human history. The Eocene epoch in which Ida lived was a critical turning point in evolution because the creatures with whom we share the planet were emerging then. And the fact that Ida is 95 percent complete eliminates the debatable statistical analysis that is often formulated out of incomplete skeletons and crushed bones.

"We are setting the bar much higher by saying this is a complete specimen, look what you can learn from it," Hurum says. Hurum likens Ida to *Archaeopteryx*. Beginning in 1860, the discoveries of different *Archaeopteryx* fossils have led scientists to conclude that the 150-million-year-old species is the best candidate to be a transitional fossil between dinosaurs and birds. *Archaeopteryx* has become an important part of the evolutionary chain because it is the oldest, most complete specimen of its kind. About Ida, Hurum says, "She's the first link in the chain between early primates and us because she's so complete, so that will be the basis for our understanding of early primates for many years."

Inevitably, one of the most basic questions asked will be, Is Ida the fossil of a human ancestor?

Hurum predicts that debate over Ida will fall into three groups. The first group will claim that she is a prosimian, a lemur. Most of the previous species found from the Eocene have been classified as lemurs, so this is the simplest conclusion. The historical record on the lemur line of fossils is so fragile and incomplete, however, that there are many holes in the theories of these discoveries.

The second group, Hurum says, will claim that Ida is an anthropoid, an early branch of primates, because some of her characteristics point to this conclusion. The shape of her face, with its short snout and steep lower jaw, is very anthropoid-like. The tooth count is also close to that of a younger anthropoid, such as those found in Egypt.

The third group, to which Hurum ascribes and which he predicts will define the fossil, will say that this is the best specimen from the Eocene that science has, and that she is the closest we have to the stem of human evolution. His scientific team is the leader of this group.

Hurum recalls his own group's skepticism in declaring that Ida is an in-between species. Because so many scientists had already concluded that there were lemurs in the Messel Pit, Hurum initially thought that might be what they had.

"We started believing she was a lemur, because that was the easiest solution, but when we looked in detail at all the traits, we started to say, this doesn't fit a lemur's characteristics," Hurum recalls.

Ida doesn't have the grooming claw that all lemurs have, and there are other major differences. One of Ida's foot bones, the astragalus, is not consistent with that of a lemur. This bone is one of the main characteristics that scientists use to identify a lemur because it generally preserves so well and has a peculiar shape. Ida's heel bone, the calcaneus, also looks more like that of an anthropoid than that of a lemur. And Jens Franzen has found a further clue in the talus bone of the foot. A specimen in the Natural History Museum in Basel, Switzerland, contains the same bone and it is clearly anthropoid—further evidence that places Ida near the anthropoid lineage.

Despite the anthropoid-like shape of Ida's face and the tooth count, her teeth were not those of a true anthropoid. "The teeth certainly match her to other primates in this general family known at that time," Holly Smith explains. "Her teeth have a kind of simple pattern. It's a species that certainly isn't around today, but in terms of the developmental pattern of what teeth are developing, it tells me that it was a primate that grew up more quickly than most of the apes and monkeys around."

From an extensive study of Ida's teeth, Smith was able to determine Ida's age and her life expectancy. "She could have lived as long as twenty years, but she would have been an adult in three to four years," Smith concludes. "She was probably a third of the way to being an adult when she died."

Like her fellow scientists, Smith points out that having a nearly complete specimen to study grounds the find in tangibles. Instead of grappling to align pieces, the scientists can draw conclusions from something solid.

"Finding a juvenile always helps people understand that these are real living things," Smith continues. "It's quite important to have a whole skeleton. For example, there is an early Egyptian skeleton called *Apidium* [a part of the Fayum fossils] that's now got a lot of skeletal material, but each and every piece was recovered separately in a big quarry. You have to sort it out from any other possible primates and try and put together what one animal looked like, but yet you have males and females and individuals of different sizes all mixed up. [Ida] shows you one complete, coherent animal with true proportions."

Smith is less interested in the inevitable debate of where Ida fits into the evolutionary chain than she is about how her appearance will change scientists' examination of early primates.

"The top controversies will be about ancestry and phylogeny [historical lineage], but in many other ways, it will put people on the right track in terms of looking at the locomotion of these animals and giving us a footing into their life history, and how they grew up," Smith says.

Philip Gingerich agrees. "In a sense it raises the standard of what a good skeleton is. I think [this] will raise the level of the game because people will know it's possible to find them and they'll work harder."

The revelations that come from Ida will likely spark an even greater interest in the Messel Pit and the specimens collected there. Inevitably, there will also be a spike in price at fossil fairs for anything collected there. Whether there are private collectors with any other complete primate specimens remains to be seen.

Hurum, for one, is skeptical that such specimens will come to light. "With all this digging going on for thirty years and only one specimen found, I think we bought the last one," he says. "But you never know. The fun thing about paleontology is that you can change the textbook forever with a hammer, and it is the only science where you can do that. All other sciences have so many black boxes now. You have to use strange machines and do strange things that people do not understand. But paleontology is really easy to communicate. You find something, and it tells you something."

From Ida's clear and convincing features, Hurum and his team determined that the lemur-like and anthropoid-like characteristics were so primitive that Ida couldn't be conclusively called either one. They realized, for the first time, that they were really trying to describe an early primate from a collection of characteristics that should be present in a very early primate, and that Ida appears right at the branching point.

"All our anatomical details point in the direction that this is something more than just another lemur," Hurum says. "That makes the specimen much more important."

But as Ida begins to help scientists fill in gaps in primate evolution, a vast chasm still remains in the evolutionary record, Franzen explains. The oldest primate is believed by many to be the 58-million-year-old *Altiatlasius*, which was found in Morocco, though its taxonomic position has been debated because the specimen consisted of just ten teeth. The next clear discovery in the anthropoid line were the Fayum primates, found near Cairo, Egypt, and determined to be 35 million years old.

"These are the two lines of evidence, but that is a gap of time of about twelve million years we are dealing with, between the Messel primate and the Fayum primates," Franzen says. "We are not dealing with the first primate; it's something which is quite near to anthropoid. What we need [from Ida] is an explosion of scientific debate."

In his book *The Hunt for the Dawn Monkey*, paleontologist Chris Beard makes the case largely through statistical analysis that his assistant's discovery of a piece of jaw from the tiny prosimian *Eosimias* in central China links Asian primates with anthropoids.

"He might be right, but at the moment, it's so fragmented that it's really hard to say," Hurum says. "We need more evidence. We need to find skulls, whole skeletons, not only small pieces. He has a broken piece of a jaw and some anklebones from the same locality, but we don't know they are from the same animal. The only way you can show that is through statistics. I'm not criticizing them, because that's the material they have and they have to try to understand it, but a complete specimen might tell a different story."

Franzen believes that Ida's greatest contribution to the evolutionary debate may be that her nearly complete specimen shows evidence of a transition because it has both primitive and derived characteristics at the same time.

"We see a combination of characteristics all in one skeleton," Franzen says. "It goes together. Otherwise, scientists saying that this particular fossil and those cranial bones really blend is hypothetical. Now we have it in one skeleton. We can say this is a whole, complex character with several traits pointing in an anthropoid direction."

Gingerich points out that Ida's completeness opens up richer comparisons to future discoveries that lived before and after her.

"The advantage of having a skeleton this complete is, hopefully it will let us make the connection to what came later," Gingerich says. "In a sense, studying primate evolution is all about looking at the diversity living today and tracing that back through time and seeing when groups converge with each other, and often we find additional diversity that we wouldn't have known about if we just studied living ones. But we're interested here to see how apes and monkeys trace back; how lemurs trace back; how tarsiers trace back, and which of these, or all of them, can we find in the Eocene. To really understand what you're dealing with, you have to know something about the teeth, the skull, the forelimbs, the hind limbs, the hands, the feet, the tail, all of this together to see whether it's consistent in representing a lemur, representing a monkey—or, if it's a mixture of things, it might be a common ancestor of them."

Going forward, Hurum hopes the chain will be filled in with an earlier species on the tree of evolution.

"We need something that looks even less like a primate, something farther down the tree, looking almost like a tree shrew," Hurum says. "The next step would be to try to go even further back in time and try to understand how our ancestor looked sixty million and a hundred million years ago."

"There is so much missing," Hurum continues. "We need almost everything. We are still fumbling in the dark. We have a piece here and a piece there, but we can't see a good picture of primate evolution. There are so many hundreds of thousands of pieces needed to understand the whole puzzle."

By February 2009, Ida's international launch had been coordinated between the Oslo Natural History Museum and the scientific team. And at a dinner at the historic Statholdergaarden restaurant in Oslo, the realization of what was to come began to sink in. A relatively small museum in Oslo was quite possibly on the edge of making a landmark scientific revelation, and at one point, Franzen turned to Professor Roaldset and said, "I really think with this fossil we could rewrite history."

The plan was for Ida to appear in New York, later to return to her home at the Oslo museum, and then go to Germany for a temporary display at the new Messel visitor center.

Once back in Oslo for good, Ida would undergo further study and be on permanent display in the Natural History Museum. Having concluded the first anatomical description, the original team of scientists would study more of her functional anatomy, morphology, and pathology in order to further specify how she moved, how she grabbed, and how she used her fingers and toes. They would also compare pieces of other skeletons from the Messel Pit with her, to learn more about her life and interaction with other creatures. All of the studies would be detailed in scientific papers.

"A complete skeleton is something you can work on for the rest of your career," Hurum says. "It's something that has many questions inside of it. When it's this old, you need to compare it to many different animals living today and other fossils too. How did it move? How were all the muscles attached to the skeleton? How was the tail moving? How was the hand gripping? There are so many questions you want to try to answer. A specimen like this will probably generate twenty scientific papers."

When Hurum told his close friend Philip Currie, a notable Canadian paleontologist, in confidence about the discovery weeks before the announcement, Currie said to Hurum, "This specimen is going to follow you for many years." Hurum laughs as he considers the truer meaning of that statement— that he has years of work ahead of him on this one specimen.

Hurum hopes that future generations will continue to examine the marvels not only of Ida but also of the Eocene and Messel Pit. When Hurum settled on the name Ida as a tribute to his five-year-old daughter, he did so believing that this young creature would have a profound impact on future biology textbooks. "Pieces of crushed bone do not make the same impact as something that looks like something, not only for specialists but for people," Hurum says. "Ida will be the icon for the early evolution of primates, because she is something the layman can look at and understand is like a monkey or an ape. You are not looking at something a scientist has to explain. You can draw your own conclusions on Ida."

In the texture of her bones and the almost eerie reach of her fingers, it's easy to see something at once alien and recognizably familiar—an animal that predates the oldest fossil of a walking hominid by 44 million years yet clearly has her own personality and character. Ida is such a complete skeleton, and such a unique find, that anyone can look at her bones and see not only into the Eocene of her birth but much deeper too, into the life that just might have given us our own.

EPILOGUE

There will be readers who will find it somewhat strange that the author refers to the apes and ourselves collectively as *we*, as if *Homo sapiens* were simply another ape. Is it not strange indeed to see Ida, a 47-million-year-old creature who was scarcely even anthropoid, as an ancestor, worthy to take her place in the family album, at least among the great-aunts? Surely we are above these creatures. Surely it is blasphemy to pretend otherwise.

Many have thought so. Many a philosopher and cleric have condemned biologists for daring to emphasize our affinity with other creatures. Many biologists—including some paleoanthropologists—have felt the same way. Many objected almost violently to the findings of Sarich and Wilson when they announced, in the 1960s, that humans and chimps shared a common ancestor just a few million years ago. Beneath their scientific objections (so some sociologists of science have observed) lay a subconscious desire to put as much distance as possible between ourselves and the apes. A common ancestor from the Early Miocene or preferably the Oligocene would be quite recent enough.

But many a philosopher and cleric, and of course many a biologist, have not been ashamed to be associated with the other beasts. St. Francis of Assisi, felt by many to be the most Christlike of the Christian saints, declared that the animals and plants were his brothers and sisters. Charles Darwin, who suggested that all creatures must have arisen in the deep past from a common ancestor, said in effect that this is literally the case. If everything is God's Creation, why would we want to be aloof from it? Who are we to be so superior?

This isn't just a point of theological whimsy. Our sense of affinity with nature as a whole—or our lack of it—profoundly affects our attitude toward it. Attitude is all: it affects the way we live our lives day to day, and it affects all politics and economics. Our sense of aloofness has led us to devise ways of living and an economy that seems designed to emphasize our separateness from nature. There is little or no conception in the modern Western economy—or in much of modern Western science—that we are a part of nature. We seem to take it for granted that the world is ours to exploit, at will, and at whim. If other species (or, indeed, other peoples) get wiped out in the process, well, that is too bad. If we truly felt, as St. Francis did and Darwin's ideas suggest we should, that other creatures are our kin—distant relatives perhaps but kin nonetheless—then surely we would not treat them simply as commodities, or sweep them aside as an inconvenience. If we did not exploit our fellow creatures so insouciantly and trash their habitats, it would be good for us too. Humanity is in a very perilous state largely because we have made such a mess of the world. To a significant extent, a sense of kinship with other creatures would be enlightened self-interest.

There are significant points of detail too. Scientists have

been warning of climate change, and spelling out the reasons for it, for at least two decades. The world's most powerful governments have begun to take an interest, up to a point, only very recently, as major Western cities such as New Orleans have come into nature's firing line. There are surely many reasons for their inertia, but one very clearly was—and is—incredulity. Politicians are not generally aware of Earth sciences and simply cannot believe that the world could be so radically different from the way it is now. New York is hot in summer and cold in winter, and that's the way the world is. New York is made of concrete and steel and built on bedrock and should last forever. To suggest that it could be buried under half a mile of ice or submerged under more than twenty feet (6 to 7 meters) of subtropical ocean is to politicians not versed in Earth sciences obvious fantasy. But it is not fantasy. It is fantastical, to be sure, and yet it is the case. If we have a sense of history—not of the past few years, or decades, or even mere centuries, but of the long history of humanity and the Earth—then we see that this is eminently possible. One day there will be another Ice Age, and some of our biggest cities will again be under ice, just as they have been in the past. For the time being, though, we are heading for a world that is more like the Eocene, and many a coastal city will simply disappear beneath the rising seas, and other places that now are wet will be desert, and some northern cities will be swallowed up by jungle as the Mayans' were, and so on and so forth. It is all eminently credible if we have a sense of the deep past. The deep past needs to be taken very seriously indeed.

In so many ways, we have so much to learn from Ida and her world.

ACKNOWLEDGMENTS

I did not write every word in this book. I wrote the bits about the science—chapters 3 to 8, plus the epilogue—and Josh Young wrote the rest, about the Oslo team and the background to the research. It has been good fun, and I am very grateful to the people who invited me to do it and smoothed the path.

Thanks, then, to Anthony Geffen of Atlantic Productions and his team: Lucie Ridout, Melissa Blanch, and Kelly Nobay. For assistance with the images that appear in this book, thanks to the Hessisches Landesmuseum Darmstadt, the Senckenberg Research Institute, and photographer Sam Peach. Thanks too to John Parsley, editor at Little, Brown, who seemed to remain remarkably unfazed by the stress, and to Pamela Marshall, the book's copyeditor. And thanks to Jonathan Lloyd, Claudia Perkins, Tim Walker, Jay Hunt, Carol Sennett, Alexander Hesse, Abbe Raven, Nancy Dubuc, David McKillop, and Dirk Hoogstra, without whose help this book might never have happened. Also to Jørn Hurum, leader of the Oslo team, who gave prompt and clear answers to all my inquiries, along with fellow scientists Jens Franzen, Jörg Habersetzer, and Holly Smith.

ACKNOWLEDGMENTS

I am very grateful too to all the people who have taught me biology over the past half century or so—and in this context particularly those who have enhanced my appreciation of primates and primate evolution. Roger Lewin, who wrote several pioneering books on human evolution, was generous with his ideas and contacts when we were both connected with *New Scientist*. In BBC days, in the 1980s, Alison Wood (née Richards) first got me seriously involved when she invited me to South Africa to help make a radio program based on the diamond jubilee symposium to celebrate Raymond Dart's discovery of *Australopithecus africanus*. I learned a great deal on that trip, especially from long conversations with Bernard Wood, then a professor at Liverpool University and now Professor of Human Origins at George Washington University; and also from Phillip Tobias, codiscoverer of *Homo habilis*, professor at the University of Witwatersrand, Johannesburg. Back in England, Sarah Bunney, who edited the *Cambridge Encyclopedia of Human Evolution*, has given me plenty of good advice on matters primatological. I continue to enjoy excellent conversations with Oliver Curry, now carrying out behavioral research in Oxford; and Jennifer Scott, who did excellent work on lowland gorillas and is now at Yale University. I first met both of them in the early 1990s when we were all connected in varying degrees to the London School of Economics (center of excellent studies on matters Darwinian).

Thanks too to my agents, Felicity Bryan, Catherine Clarke, and Michele Topham. But most of all I am grateful to my wife, Ruth, without whose encouragement and organizational skills I doubt I would get anything done at all.

INDEX

251

INDEX

apes and humans, 122–23, 155, 186–87, 190, 197. *See also* humans
Apidium moustafai (primates), 180, 238
Araceae family, 71
Archaeolemur (lemur), 117
Archaeopteryx (bird), 47, 50–52, 68, 161, 163, 216, 236
archaic forms, 199. *See also* primates
Arctic Circle, 38, 43, 231
Arctic Ocean, 11, 54, 56, 58, 167
Ardipithecus ramidus (hominid), 196–97
artificial matrix, 23, 25, 32
artiodactyls, 22, 47, 98
"Ascent of Man," 16
Asia, 55, 58, 122, 166
    fossils from, 22, 156, 173–75, 181–82, 184–87, 214–15; importance of, 177–78
    "lesser" and "great" apes of, 183
    Southeast, 71, 166; animals still living in, 96, 119, 150
    See also China; India; Indonesia
Asiatosuchus genus, 82
asteroid theory, 39–40, 45, 145–46, 161
Atlantic Ocean, 51, 72
Atlantic Productions (London), 231
atomic weights, 41
Atractosteus strausi (fish), 77
Attenborough, Sir David, 133, 234
Australia, 92, 166–67
Australopithecus africanus, A. afarensis, A. habilis (hominids), 176, 194–96, 197
aye-aye, 91, 116
Azolla (aquatic fern), 52–54, 56, 167, 169
Azolla Event, 54, 167

Baryphracta deponiae (crocodile), 83
Basilosaurus (whale), 47
bats, 34, 49–50, 64, 94, 98–99, 104
Beamon, Bob, 121
Beard, Chris, 214, 240
Bearder, Simon, 119
Beatles, The, 195
Belgium, fossils from, 152, 153
Berger, Lee, 235
Beringian land bridge, 173–74
"big bang," 39

biostratinomy, 26
birds, 37, 50, 64, 83–87
    age of, 145
    descendants of dinosaurs, 45, 48, 85, 145, 236
    flightless, 48–49, 84–85, 117–18
    nectar feeders or insectivorous, 86, 91
    predators, 87, 128–29; megapredators, 48–49
    resemblance to modern, 37, 47, 83
    as toolmakers, 190
    waterfowl, 84, 86; Sun Bittern, 84
    See also Archaeopteryx
brains, 43–44
    hominid, 197–98; human vs. animal, 190–95
    primate, 124–30, 131, 181; archaic, 153
Brown, Peter, 235
Brunet, Michel, 234
bush babies, 101, 104, 106, 113, 118–21, 174, 205, 224–25
Buxolestes (otterlike mammal), 89–90, 99

calcaneus, elongated, 119–20, 121, 213
Canada, 38, 58, 177
Cantius (prosimian), 154
carbon, atomic weights of (12–14), 41
carbon dioxide. *See* CO_2
carbonic acid rain, 55
Carnegie Museum of Natural History, 214
Carnivora order and carnivores, 88, 91, 93
    emergence of, 46–47, 48, 88
cassowary, 48–49
catarrhines (Old World monkeys), 104, 125, 181, 183–84, 220
    evolution of, 155–57, 207–8; split into two groups, 122, 156, 179–80
"catastrophism," 161–62
Catopithecus (anthropoid), 181
Cenozoic era, 38, 39, 165
Central America, 80, 96
Cercamonius (primate), 222
CH_4 levels. *See* methane levels
Chad: early fossil found in, 197, 235
chalicotheres, 105

INDEX

Cheirogaleidae family, 117
chimpanzees, 184–86, 189–91, 194
  related to humans, 123, 140–41, 175,
    234, 245; divergence from, 176,
    183, 187, 192
China, 82
  fossils from, 177, 185, 214–15, 240;
    "dragon teeth," 188; Peking Man,
    110, 198
Chiroptera order, 104
Chomsky, Noam, 191
clades, 110–13, 122, 205, 207
climate, 380. *See also* Eocene epoch;
  rain forest; temperatures
climate change, 7, 97, 146, 247. *See also*
  global warming
clingers and leapers, 120–22, 154
  Ida as, 27, 122, 220, 224, 229
CO₂ (carbon dioxide), 56, 96–97, 172
  anesthetic effect of, 34, 50, 63, 99, 226
  combines with metals or rocks,
    33–34, 54–55, 65
  as greenhouse gas, 40, 42, 57, 168;
    Azolla and, 167
Colocasia (taro), 71. *See also* plants,
  flowering and nonflowering
Colorado, fossils from, 151
convergence, 75, 83, 91, 93, 211
  idea rejected, 208
contact microradiography, 24
continental drift, 7, 50–51, 55, 57–59,
  152, 166–67, 177
  land bridges, 58–59, 72–73, 158,
    166, 173–74
Cook, Captain James, 150
coprolites (fossil feces), 50, 99
Coraciiformes, 84. *See also* birds
crater lake. *See* maar (crater) lake
creodonts, 46, 91–92
Cretaceous period, 39, 142, 145, 152
  mammals of, 44–45, 98, 146–47;
    primates or primate ancestors,
    148, 150–51
  plants of, 43; grass, 170
  reptiles of, 79, 81
crocodiles. *See* reptiles
Cro-Magnons, 201
Cryptodira genus, 81
CT scans, 21, 24, 26, 31, 33, 233
Currie, Philip, 243

Cuvier, Georges, 216
cyanobacteria, 52–53
Cyclurus kehleri (fish), 77–78
Cyperaceae family, 71

Dactylopsila (marsupial), 91
Darmstadt Museum, 25, 50, 67–68, 69
Dart, Raymond, 194–96
Darwin, Charles, 16, 44, 75, 137, 150,
  158, 175, 246
  criticism of, 159–62, 216
  Ida dedicated to, 222, 230
  post-Darwinism, 143, 148, 159,
    192–93, 195
Daubentoniidae family, 116
"Dawn Ape." *See Aegyptopithecus*
"Dawn Monkey." *See Eosimias*
Dawn Monkey, The Hunt for the
  (Beard), 240
Dawson, Charles, 193
Descartes, René, 190–91
Diatryma (flightless bird), 48–49, 85
dinosaurs, 11, 98
  coining of term, 159
  birds as descendants of, 45, 48, 85,
    145, 236
  extinction of, 39–40, 42–46,
    143–46, 161–62
  feathered, 215
  mammals and, 13, 43–44, 142–48;
    humans, 147
Diplocynodon darwini (crocodile), 66, 82
divergence, 75, 83, 157, 183, 187
DNA studies, 111, 119, 123, 140,
  182–83
Dobzhansky, Theodosius, 57
Doliostrobus (extinct conifer), 71
Drake Passage, 51
Dryopithecus (anthropoid), 185
Duke University, 178

eagles, and Monkey-Eating eagle, 48
ecosystems, 5–6, 73, 98
eels, 78–79
Egypt, 22, 155, 175, 182, 227. *See also*
  Fayum Depression
elephants, 47, 73, 108, 139, 166
  dwarf, 199
Ellis, William, 150
Emys (turtle), 81

INDEX

INDEX

INDEX

INDEX

INDEX

INDEX

INDEX

INDEX